Design of
Thermal Systems

Design of Thermal Systems

W. F. Stoecker

Professor of Mechanical Engineering
University of Illinois at Urbana-Champaign

TEST #2 → CH. 4-7 ON MONDAY MAR.21

McGraw-Hill Book Company
New York   St. Louis   San Francisco
Düsseldorf   Johannesburg   Kuala Lumpur
London   Mexico   Montreal
New Delhi   Panama   Rio de Janeiro
Singapore   Sydney   Toronto

Design of Thermal Systems

Library of Congress Catalog Card Number 75-141927

07-061617-5

7 8 9 10   KPKP   7832109

Contents

Preface

The title, "Design of Thermal Systems," reflects the three concepts embodied in this book: *design, thermal,* and *systems.*

Design A frequent product of the engineer's efforts is a drawing, a set of calculations, or a report that is an abstraction and description of hardware. Within engineering education, the cookbook approach to design, often practiced during the 1940s, discredited the design effort so that many engineering schools dropped design courses from their curricula in the 1950s. But now design has returned. This reemergence is not a relapse to the earlier procedures; design is reappearing as a creative and highly technical activity.

Thermal Within many mechanical engineering curricula the term *design* is limited to *machine design*. In order to compensate for this frequent lack of recognition of thermal design, some special emphasis on this subject for the next few years is warranted. The designation *thermal* implies calculations and activities based on principles of thermodynamics, heat transfer, and fluid mechanics.

The hardware associated with thermal systems includes fans, pumps, compressors, engines, expanders, turbines, heat and mass exchangers, and reactors, all interconnected with some form of conduits. Generally, the working substances are fluids. These types of systems appear in such industries as power generation, electric and gas utilities, refrigeration, air conditioning and heating, and in the food, chemical, and process industries.

Systems Engineering education is predominantly *process oriented*, while engineering practice is predominantly *system oriented*. Most courses of study in engineering provide the student with an effective exposure to such processes as the flow of a compressible fluid through a nozzle and the behavior of hydrodynamic and thermal boundary layers at solid surfaces. The practicing engineer, however, is likely to be confronted with a task such as designing an economic system that receives natural gas from a pipeline and stores it underground for later usage. There is a big gap between knowledge of individual processes and the integration of these processes in an engineering enterprise.

Closing the gap should not be accomplished by diminishing the emphasis on processes. A faulty knowledge of fundamentals may result in subsequent failure of the system. But within a university environment, it is beneficial for future engineers to begin thinking in terms of systems. Another reason for more emphasis on systems in the university environment, in addition to influencing the thought patterns of students, is that there are some techniques—such as simulation and optimization—which only recently have been applied to thermal systems. These are useful tools and the graduate should have some facility with them.

While the availability of procedures of simulation and optimization is not a new situation, the practical application of these procedures has only recently become widespread because of the availability of the digital computer. Heretofore, the limitation of time did not permit hand calculations, for example, of an optimization of a function that was dependent upon dozens or hundreds of independent variables. This meant that, in designing systems consisting of dozens or hundreds of components, the goal of achieving a *workable* system was a significant accomplishment and the objective of designing an *optimum* system was usually abandoned. The possibility of optimization represents one of the new facets of design.

Outline of this book The goal of this book is the design of optimum thermal systems. Chapters 6 through 11 cover topics and specific procedures in optimization. After Chap. 6 explains the typical statement of the optimization problem and illustrates how this statement derives from the physical situation, the chapters that follow explore optimization procedures such as calculus methods, search methods, geometric programming, dynamic programming, and linear programming. All these methods have applicability to many other types of problems besides thermal ones and, in this sense, are general. On the other hand, the applications are chosen from the thermal field to emphasize the opportunity for optimization in this class of problems.

If the engineer immediately sets out to try to optimize a moderately complex thermal system, he is soon struck by the need for predicting the performance of that system, given certain input conditions and performance characteristics of components. This is the process of *system simulation*. System simulation not only may be a step in the optimization process but may have a usefulness in its own right. A system may be designed on the basis of some maximum load condition but may operate 95 percent of the time at less-than-maximum load. System simulation permits an examination of the operating conditions that may pinpoint possible operating and control problems at nondesign conditions.

Since system simulation and optimization on any but the simplest problems are complex operations, the execution of the problem must be performed on a computer. When using a computer, the equation form of representation of the performance of components and expression of properties of substances is much more convenient than tabular or graphical representations. Chapter 4 on mathematical modeling presents some techniques for equation development for the case where there is and also where there is not some insight into the relationships based in thermal laws.

Chapter 3, on economics, is appropriate because engineering design and economics are inseparable, and because a frequent criterion for optimization is the economic one. Chapter 2, on workable systems, attempts to convey one simple but important distinction—the difference between the design process that results in a workable system in contrast to an optimum system. The first chapter on engineering design emphasizes the importance of design in an engineering undertaking.

The appendix includes some problem statements of several comprehensive projects which may run as part-time assignments during an entire term. These term projects are industrially oriented but require application of some of the topics explained in the text.

The audience for which this book was written includes senior or first-year graduate students in mechanical or chemical engineering, or practicing engineers in the thermal field. The background assumed is a knowledge of thermodynamics, heat transfer, fluid mechanics, and an awareness of the performance characteristics of such thermal equipment as heat exchangers, pumps, and compressors. The now generally accepted facility of engineers to do basic digital computer programming is also a requirement.

Acknowledgments   Thermal system design is gradually emerging as an identifiable discipline.   Special recognition should be given to the program coordinated by the University of Michigan on Computers in Engineering Design Education, which in 1966 clearly delineated topics and defined directions that have since proved to be productive.   Acknowledgment should be given to activities within the chemical engineering field for developments that are closely related, and in some cases identical, to those in the thermal stem of mechanical engineering.

Many faculty members during the past five years have arrived, often independently, at the same conclusion as the author: the time is opportune for developments in thermal design.   Many of these faculty members have shared some of their experiences in the thermal design section of *Mechanical Engineering News* and have, thus, directly and indirectly contributed to ideas expressed in this book.

This manuscript is the third edition of text material used in the Design of Thermal Systems course at the University of Illinois at Urbana-Champaign.   I thank the students who have worked with me in this course for their suggestions for improvement of the manuscript.   The second edition was an attractively printed booklet prepared by my Department Publication Office, George Morris, Director; June Kempka and Dianne Merridith, typists; and Don Anderson, Bruce Breckenfeld, and Paul Stoecker, draftsmen. Special thanks are due to the Engineering Department of Amoco Chemicals Corporation, Chicago, for their interest in engineering education and for their concrete evidence of this interest shown by printing the second edition.

Competent colleagues are invaluable as sounding boards for ideas and as contributors of ideas of their own.   Professor L. E. Doyle offered suggestions on the economics chapter and Prof. C. O. Pedersen, a coworker in the development of the thermal systems program at the University of Illinois at Urbana-Champaign, provided advice at many stages.   Mr. Donald R. Witt and a class of architectural engineering students at Pennsylvania State University class-tested the manuscript and provided valuable suggestions from the point of view of a user of the book. Beneficial comments and criticisms also came from the Newark College of Engineering, where Prof. Eugene Stamper and a group of students tested the manuscript in one of their classes. Professor Jack P. Holman of Southern Methodist University, consulting editor of McGraw-Hill Book Company, supplied perceptive comments both in terms of pedagogy as well as in the technical features of thermal systems.

The illustrations in this book were prepared by George Morris of Champaign, Illinois.

By being the people that they are, my wife Pat and children Paul, Janet, and Anita have made the work on this book, as well as anything else that I do, seem worthwhile.

W. F. Stoecker

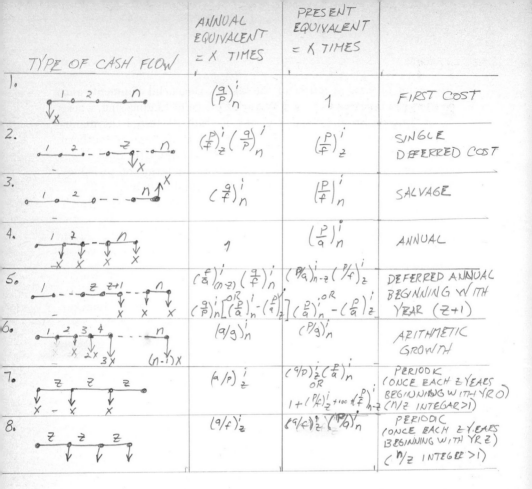

TYPE OF CASH FLOW	ANNUAL EQUIVALENT = X TIMES	PRESENT EQUIVALENT = X TIMES	
1.	$\left(\frac{a}{p}\right)^i_n$	1	FIRST COST
2.	$\left(\frac{p}{f}\right)^i_z\left(\frac{a}{p}\right)^i_n$	$\left(\frac{p}{f}\right)^i_z$	SINGLE DEFERRED COST
3.	$\left(\frac{a}{f}\right)^i_n$	$\left(\frac{p}{f}\right)^i_n$	SALVAGE
4.	1	$\left(\frac{p}{a}\right)^i_n$	ANNUAL
5.	$\left(\frac{f}{a}\right)^i_{(n-z)}\left(\frac{a}{f}\right)^i_n$ OR $\left(\frac{a}{p}\right)^i_n\left[\left(\frac{p}{a}\right)^i_n-\left(\frac{p}{a}\right)^i_z\right]$	$\left(\frac{p}{a}\right)^i_{n-z}\left(\frac{p}{f}\right)^i_z$ OR $\left(\frac{p}{a}\right)^i_n-\left(\frac{p}{a}\right)^i_z$	DEFERRED ANNUAL BEGINNING WITH YEAR (z+1)
6.	$\left(\frac{a}{g}\right)^i_n$	$\left(\frac{p}{g}\right)^i_n$	ARITHMETIC GROWTH
7.	$\left(\frac{a}{p}\right)^i_z$	$\left(\frac{a}{p}\right)^i_z\left(\frac{p}{a}\right)^i_n$ OR $1+\left(\frac{p}{f}\right)^i_z+...+\left(\frac{p}{f}\right)^i_{n-z}$	PERIODIC (ONCE EACH z YEARS BEGINNING WITH YR 0) (n/z INTEGER >1)
8.	$\left(\frac{a}{f}\right)^i_z$	$\left(\frac{a}{f}\right)^i_z\left(\frac{p}{a}\right)^i_n$	PERIODIC (ONCE EACH z YEARS BEGINNING WITH YR z) (n/z INTEGER >1)

1
Engineering Design

1-1 Introduction Some typical professional activities of engineers are sales, construction, research, development, and design. The last-named activity—design—will be our special concern in this book. The immediate product of the design process is a report, a set of calculations, or a drawing that is an abstraction of hardware. The subject of the design may be a process, an element or component of a larger assembly, or an entire system.

Our emphasis will be on *system* design, wherein a system is defined as a collection of components with interrelated performance. Even this definition often needs interpretation, because a large system sometimes includes subsystems. Furthermore, we shall progressively focus on *thermal* systems where fluids and energy in the form of heat and work are conveyed and converted. Before adjusting this focus, however, this chapter will examine the larger picture into which the technical engineering activity blends. We shall call this larger operation an engineering *undertaking* which implies that engineering plays a decisive role but also dovetails with other considerations. Engineering undertakings include a wide variety of commercial and industrial enterprises as well as municipal-, state-, and federally sponsored projects.

1-2 Decisions in an engineering undertaking In the past 10 years an appreciable amount of attention has been devoted to the "methodology" or the "morphology" of engineering undertakings. Studies on these topics have analyzed the steps and procedures used in reaching decisions. One contribution of these studies has been to stimulate the engineer to reflect on his own thinking processes and on that of others on the project team. Certainly the process and sequence of steps followed in each undertaking is different, and no one sequence, including the one described in this chapter, is universally applicable. The starting point, the goal, and the side conditions differ from one undertaking to the next, so the procedures will vary.

The advantage of analyzing the decision process, especially in complex undertakings, is that it leads to a more logical coordination of the many individual efforts that comprise the entire venture. The flow diagram in Fig. 1-1 shows the typical steps followed in the conception, evaluation, and execution of the plan. The rectangular boxes, which indicate actions, may represent considerable effort and expenditures on large projects. The diamond boxes represent decisions, such as whether to continue the project or to drop it.

The technical engineering occurs mostly in activities 5 and 7—product or system design—and in research and development. Very little will be mentioned in this chapter about product or system design, because it is the subject of the entire study of thermal systems that will be studied in the chapters to follow. The flow diagram only shows how this design procedure fits into the larger pattern of the undertaking. The individual nondesign activities will be discussed next.

1-3 Need or opportunity (step 1) Step 1 in the flow diagram of Fig. 1-1 is to define the need or opportunity. It may seem to be a simple task to state the need or opportunity, but such is not always the case. For example, the officials of a city may think that their need is to enlarge a reservoir for storage of a larger quantity of water for municipal purposes. The officials may not truly have specified the need, but instead have leaped to one possible solution. Perhaps the need would best have been stated as a low-water reserve during certain times of the year. Enlargement of the

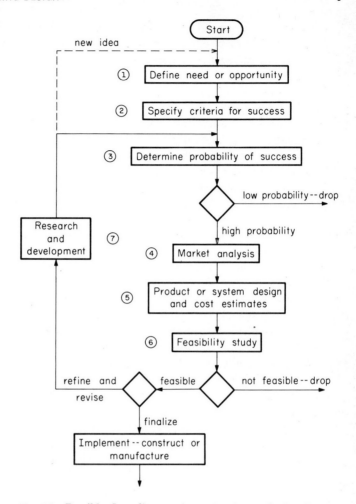

Fig. 1-1 Possible flow diagram in evaluating and planning an engineering undertaking.

reservoir might be one possible solution to consider, but other solutions might be to restrict the consumption of water or to seek other sources such as wells. Sometimes, possible solutions are precluded by not starting at the proper statement of the need.

The word "opportunity" has positive connotations, whereas "need" suggests a defensive action. Sometimes the two cannot be distinguished. For example, an industrial firm may recognize a new product as an opportunity, but if the company does not then expand its line of products, business is likely to decline. The introduction of a new product is, thus, also a need.

In commercial enterprises, typical needs or opportunities lie in the renovation or expansion of facilities to manufacture or distribute a current product. Another form in which an opportunity arises is when the sale of a product not manufactured by the firm is rising and the market potential seems favorable. Still a third form in which an opportunity arises is through research and development within the organization. A new product may be developed intentionally or accidentally. Sometimes a new use can be found by slight modification of an existing product. An organization may know how to manufacture a gummy, sticky substance and assign to the research and development department the task of finding some use for this substance.

Of interest to us at the moment is the need or opportunity that requires engineering design at a subsequent stage.

1-4 Criteria of success (step 2) In commercial enterprises, the usual criterion of success is showing a profit or, more specifically, providing a certain rate of return on the investment. Also, in public projects, the criterion of success is the degree to which the need is satisfied in relation to the cost—monetary or otherwise.

In a profit-and-loss economy, the expected earning power of a proposed commercial project is a dominating influence on the decision to proceed with the project. Strict monetary concerns are always tempered, however, by human, social, and political considerations to a greater or lesser degree. Another way of stating this influence is to say that a price tag is placed on the nonmonetary factors. A factory may be located at a more remote site at a penalty on transportation costs in order that its atmospheric pollution or noise affect fewer people. As an alternate, the plant may spend a large sum of money for superior pollution control in order to be a good neighbor to the surrounding community.

Sometimes a firm will design and manufacture a product that offers little opportunity for profit simply to round out a line of products. The availability of this product, product A, permits the sales force to say to a prospective customer, "Yes, we can sell you product A, but we recommend product B," which is a more profitable item in the company's line and may actually be superior to product A.

Often a decision, particularly of an emergency nature, appears out of the realm of economics. If a boiler providing steam for heating a rental office building fails, the decision whether to repair or replace the boiler may seem to be out of the realm of economics. The question could still be considered an economic one, however, with the penalty for not executing the project being an overpowering loss.

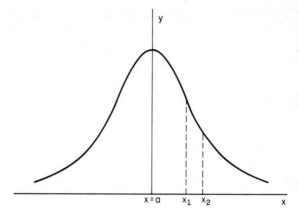

Fig. 1-2 Probability distribution curve.

1-5 Probability of success (step 3) The economic success of engineering projects is not one of the two traditional certainties of life. Along with the opportunity of achieving a profit goes the threat of loss. If the possible construction of a plant to manufacture a product is being considered, there is no absolute assurance that the plant will meet the success criteria discussed in Sec. 1-4. It is preferable, then, to speak of the likelihood or the probability of success, as listed in step 3 of the flow diagram of Fig. 1-1.

The mention of the term *probability* suggests a probability distribution curve, and this curve, as shown in Fig. 1-2, is not without applicability to the decision-making process. The significance of the distribution curve lies particularly in the evaluation of the area beneath the curve. The ordinate y indicates the probability of the event occurring between x and $x + dx$. The area under the curve between x_1 and x_2, for example, represents the probability P of the event occurring between values of x_1 and x_2. Thus,

$$P = \int_{x_1}^{x_2} y \, dx$$

The probability of the event occurring somewhere in the range of x is unity, so the integration over the entire range of x is equal to 1.0:

$$\int_{-\infty}^{\infty} y \, dx = 1$$

The equation for the probability distribution curve is

$$y = \frac{h}{\sqrt{\pi}} e^{-h^2(x-a)^2} \tag{1-1}$$

The maximum value of the ordinate is $h/\sqrt{\pi}$. This fact suggests that increasing the value of h alters the shape of the distribution curve, as shown in Fig. 1-3. If h_1 is greater than h_2, the peak of the h_1 curve rises higher than that of the h_2 curve.

To extend the probability idea to decision making in an engineering undertaking, suppose that a new product or facility is proposed, and the criteria for success is a 10 percent rate of return on the investment for a 5-year life of the plant. After a preliminary design, the probability distribution curve is shown as indicated in Fig. 1-4. Rough figures were used throughout the evaluation, so the distribution curve is flat, indicating no great confidence in an expected percent of return of investment of, let us say, 18 percent. The expected rate of return is attractive enough, however, to proceed further to a complete design, including cost estimates. If the most probable return on investment after this complete design were 16 percent, for example, the confidence in this figure would be greater than the confidence in the 18 percent figure after the preliminary design, because costs have now been analyzed more carefully and marketing studies have been conducted more thoroughly.

The probability distribution curves at two other stages—after construction and after 1 year of operation—show progressively greater degrees of confidence in the rate of return after a 5-year life. After 5 years, the rate of return is known exactly, and the proba-

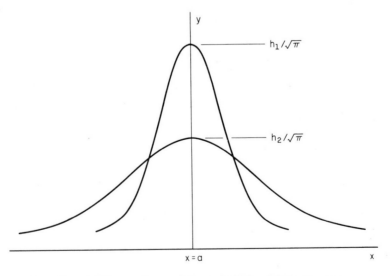

Fig. 1-3 Several different shapes of the probability distribution curve.

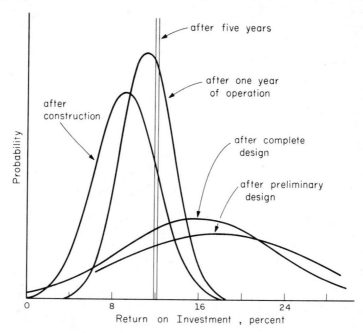

Fig. 1-4 Distribution curves at various stages of decision making.

bility distribution curve degenerates into a curve that is infinitesimally thin and infinitely high.

While decisions made on the basis of probabilities are facts of
life, the determination of the probability distribution curves cannot
be accomplished with anywhere near the accuracy suggested by Fig.
1-4. To a considerable extent the go-or-no-go decisions are made
by management on the basis of "highly likely" or "unlikely" chances
of success. The emergence of new developments that would compete with the proposed product, for example, cannot always be
anticipated. Marketing surveys are imprecise. Nevertheless,
efforts toward quantifying the probability calculation are highly
desirable.

1-6   Market analysis (step 4)   If the undertaking is one in which a
product or service must eventually be sold or leased to customers,
there must be some indication of favorable reaction by the potential
consumer. An ideal form of the information provided by a market
analysis would be a set of curves as shown in Fig. 1-5. With an
increase in price, the potential volume of sale decreases until such a
high price is reached that no sales can be made. The sales volume-

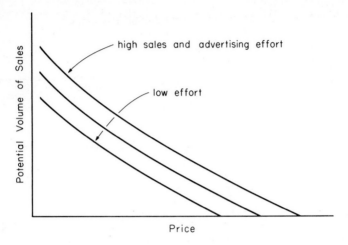

Fig. 1-5   End result of a market analysis.

price relationship affects the size of the plant or process because the unit price is often lower in a large plant.   For this reason, the marketing and plant capabilities must be evaluated in conjunction with one another.

The sales and advertising effort influence the volume of sales for a given price, so a family of curves would be expected.   Since a cost is associated with the sales and advertising effort, and since a continuous increase of this effort results in diminishing improvement in sales, there exists an optimum level of sales and advertising effort.   Simultaneously with the technical plans for the undertaking, there should emerge a marketing plan as well.

1-7   Feasibility (step 6)   The feasibility study, step 6, and the subsequent feasibility decision refer to whether the project is even possible.   A project may be feasible, or possible, but may not be economical.   Examples of infeasibility occur due to unavailability of investment capital, land, labor, or favorable zoning regulations. - Safety codes or other regulatory laws may prohibit the enterprise. If an undertaking is shown to be infeasible, either alternates must be found to circumvent the barrier, or the project must be dropped.

1-8   Research and development (step 7)   If the product or process is a new one to the experience of the organization, the results from research and development (R&D) may be an important input to the decision process.   Research efforts may provide the origin or improvement of the basic idea, and development work may provide working models or a pilot plant, depending upon the nature of the undertaking.

Placing research and development in a late stage of decision making, as was done in Fig. 1-1, suggests that an idea originates somewhere else in the organization, or in the field, and eventually is placed at the doorstep of R&D for transformation into a workable idea. The possibility of the idea originating in the research group should also be exploited and is indicated by the dashed line in Fig. 1-1. Often research people learn of new ideas in other fields which might possibly be applied in the sphere of their own activity.

1-9 Iterations The loop in Fig. 1-1 emphasizes that the decision-making process involves numerous iterations. Each pass through the loop improves the amount and the quality of information and data. Eventually, a point is reached where decisions are finalized regarding the design, production, and marketing of the product. The substance that flows through this flow diagram is information. This information may be in the form of reports and conversations and may be both verbal and pictorial. The iterations are accomplished by communication between people, with this communication interspersed by go-or-no-go decisions.

1-10 Optimization of operation The flow diagram of Fig. 1-1 terminates with the construction or beginning of manufacture of a product or service. Actually another stage takes over at this point which seeks to optimize the operation of a given facility. The facility was designed on the basis of certain design parameters which almost inevitably change by the time the facility is in operation. The next challenge, then, is to operate the facility in the best possible manner in the light of such factors as costs and prices that actually exist. A painful activity occurs when the project is not profitable and the objective becomes that of minimizing the loss.

1-11 Technical design Step 5 in Fig. 1-1, the product or system design, has so far been omitted in the discussion. The reason for this omission is that the system design is the subject of our study from this point on. This step is where the largest portion of engineering time is devoted. System design as an activity lies somewhere between the study and analysis of individual processes or components and the larger decisions, which are heavily economic. Usually one person coordinates the planning of the undertaking. This manager normally emerges with a background gained from experience in one of the subactivities. His experience might be in finance, engineering, or marketing, for example. Whatever his original discipline, the manager must become conversant with all of the fields that play a role in the decision-making process.

The word "design" encompasses a wide range of activities. Design may be applied to the act of selecting a single member or part, such as the size of a tube in a heat exchanger; to a larger component, such as the entire shell-and-tube heat exchanger; or to the design of the system in which the heat exchanger is only one component. Design activities can be directed toward mechanical devices which incorporate linkages, gears, and other moving solid members, electrical or electronic systems, thermal systems, and a multitude of others. Our concentration will be on thermal systems which find their examples in power generation, heating and refrigeration plants, the food processing industry, and in the chemical and process industries.

1-12 Summary The flow diagram and description of the decision processes discussed in this chapter are highly simplified and are not sacred. Since almost every undertaking is different, there are almost infinite variations in starting points, goals, and intervening circumstances. The purpose of the study is to emphasize the advantage of systematic planning. Certain functions are common in the evaluation and planning of undertakings, and these are particularly the iterations and the decisions that occur at various stages.

SOME INTRODUCTORY BOOKS ON ENGINEERING DESIGN

Alger, J. R. M., and C. V. Hays: "Creative Synthesis in Design," Prentice-Hall, Inc., Englewood Cliffs, N.J., 1964.

Asimow, M.: "Introduction to Design," Prentice-Hall, Inc., Englewood Cliffs, N.J., 1962.

Beakley, G. C., and H. W. Leach: "Engineering, An Introduction to a Creative Profession," The Macmillan Company, New York, 1967.

Buhl, H. R.: "Creative Engineering Design," Iowa State University Press, Ames, Iowa, 1960.

Dixon, J. R.: "Design Engineering: Inventiveness, Analysis and Decision Making," McGraw-Hill Book Company, New York, 1966.

Harrisberger, L.: "Engineersmanship, A Philosophy of Design," Brooks/Cole Publishing Company, Belmont, Calif., 1966.

Krick, E. V.: "An Introduction to Engineering and Engineering Design," John Wiley & Sons, Inc., New York, 1965.

Middendorf, W. H.: "Engineering Design," Allyn and Bacon, Inc., Boston, Mass., 1968.

Woodson, T. T.: "Introduction to Engineering Design," McGraw-Hill Book Company, New York, 1966.

2
Designing a Workable System

2-1 Introduction The simple but important point to be emphasized by this chapter is the distinction between designing a "workable" system and an "optimum" system. This chapter also continues the progression from the broad concerns of an "undertaking" as described in Chap. 1 to a concentration on engineering systems and even more specifically on thermal systems.

The statement is so frequently made that "there are many possible answers to a design problem" that the idea is sometimes conveyed that all solutions are equally desirable. Actually only one solution is the optimum where the optimum is based on some defined criteria, such as cost, size, or weight. The distinction then will be made between a "workable" and an "optimum" system. It should not be suggested that a workable system is being scorned. Obviously, a workable system is infinitely preferable to a "non-workable" system. Furthermore, extensive effort in progressing from a workable toward an optimum system may not be justified due to limitations in calendar time, cost of engineering time, or even the reliability of the basic data on which the design is based. One point that will be explained in this chapter is how in the design process superior solutions may be ruled out by eliminating some system concepts too early. Another way in which superior solutions are precluded is by fixing interconnecting parameters between components and selecting the components based on these parameters in contrast to permitting these parameters to float until the optimum total system emerges.

2-2 A workable system The definition of a workable system is one that:

1. Meets the requirements of the purpose of the system (such as providing the required amount of power, heating, cooling, or fluid flow, or surrounding a space with a specified environment so that people will be comfortable or a chemical process will proceed or not proceed)
2. Will have satisfactory life and maintenance costs
3. Abides by all constraints, such as size, weight, temperatures, pressure, material properties, noise, pollution, etc.

In summary, a workable system performs the assigned task within the imposed constraints.

2-3 Steps in arriving at a workable system The two major steps in achieving a workable system are:

1. Select the concept to be used.
2. Fix whatever parameters are necessary to select the components of the system. These parameters must be chosen so that the design requirements and constraints are satisfied.

2-4 Creativity in concept selection Engineering and especially engineering design are potentially creative activities. Some typical reasons for creativity not being exercised are limitations of time for adequate exploration, discouragement by supervision or environment, and the laziness and lack of courage of the engineer himself. It is particularly in the concept selection that creativity can be exercised. Too often only one concept is ever considered—the concept that was used on the last similar job. As a standard practice, the engineer should discipline himself to review all of the alternate concepts in some manner appropriate to the scope of the project.

Old ideas that were once discarded as impractical or uneconomical should be constantly reviewed. Costs change and new devices or materials appear on the market which may make an approach successful today that was not attractive 10 years ago.

2-5 Workable vs. optimum system The distinction between the approach used in arriving at a workable system and an optimum system can be illustrated by a simple example. Suppose that pump and piping are to be selected to convey 50 gpm from one position to another position that is 700 ft removed and 25 ft higher than the original position. If the design is approached with the limited objective of achieving a workable system, the following procedure might be followed:

1. In addition to 25 ft of head required to overcome the difference in elevation, arbitrarily provide an additional 30 ft of head to compensate for friction in the 700 ft of pipe.
2. Based on the foregoing decision, select a pump which delivers 50 gpm or more against a head of 55 ft. Finally, select a pipe size from a handbook such that the head loss in 700 ft of length is 30 ft or less. A pipe size of 2 in. satisfies this requirement.

Approaching the same problem with the objective of achieving an optimum system presupposes agreement on a criterion to optimize. A frequently chosen criterion is cost—sometimes first cost only in speculative projects or the lifetime cost consisting of first-plus-lifetime pumping and maintenance costs.

In designing the optimum pump and piping system for minimum lifetime cost, the head to be developed by the pump is not immediately fixed, but left free to float. If the three major contributors to cost are (1) the first cost of the pump, (2) the first cost of the pipe, and (3) the lifetime pumping costs, these costs will vary as a function of pump head as shown in Fig. 2-1. As the pump head increases, the cost of the pump likely increases for the required flow rate of 50 gpm, because of the need for higher speed and/or larger impeller diameter. With the increase in head, the power required by the pump increases, which is reflected in a higher lifetime pumping cost. The third contributor to the total cost is the first cost of the pipe. This cost becomes enormously high as the head available to overcome friction in the pipe reduces to zero. The available head for the pipe is the pump head minus the 25-ft difference in elevation. An appropriate optimization technique can

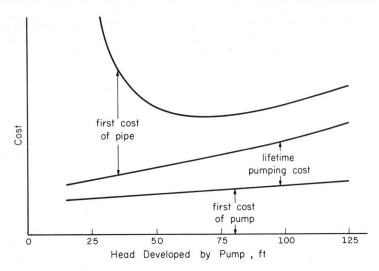

Fig. 2-1   Contributions to costs of pump and piping system.

be used to determine the optimal pump head, which in Fig. 2-1 is approximately 70 ft.   Finally the pump can be selected to develop a 70-ft discharge head, and a pipe size can be chosen such that the head loss due to friction is 45 ft or less.

The tone of the preceding discussion indicates a strong preference toward designing optimum systems.   To temper this bias, several additional considerations should be mentioned.   If the magnitude of the job is small, the cost of the increased engineering time required for the optimization may devour any saving due to optimization.   Not only engineering man-hours, but limited calendar time may not permit the design to proceed beyond a workable design.

2-6   Design of a food-freezing plant   Large-scale engineering projects are extremely complex and decisions are often intricately interrelated, not only mutually influential in the purely technical area but crossing over into the technoeconomic, social, and human fields. To illustrate a few of the decisions involved in a realistic commercial undertaking and to provide a further example of the contrast between a workable system and an optimum system, consider the following project:

1. A food company can buy sweet corn and peas from farmers during the season and sell the vegetables as frozen food throughout the year in a city 150 miles away.

2. What are the decisions and procedures involved in designing the plant to process and freeze the crops?

The statement of the task actually starts at an advanced stage in the decision process, because it is already assumed that a plant will be constructed. Realistically, this decision cannot be made until some cost data are available to evaluate the attractiveness of the project. Let us assume, therefore, that an arbitrarily selected solution has been priced out and found to be potentially profitable. We are likely, then, to arrive at a solution that is an improvement over the arbitrary selection.

Some major decisions that must be made are (1) the location, (2) the size, and (3) the type of freezing plant. The plant could be located near the producing area, in the market city, or somewhere between. The size will be strongly influenced by the market expectation. The third decision, the type of freezing plant, embraces the engineering design. These three major decisions are interrelated. For example, the location and size of plant might reasonably influence the type of system selected. The selection of the type of freezing plant includes the decision of the concept on which the freezing plant will function. After the concept has been decided, the internal design of the plant can proceed.

An outline of the sequence of tasks and decisions by which a workable design could be arrived at is as follows:

1. Decide to locate the plant in the market city adjacent to a refrigerated warehouse operated by the company.
2. Select the freezing capacity of the plant as influenced by the current availability of the crop, the potential sale in the city, and available financing.
3. Decide upon the concept to be used in the freezing plant, as for example the one shown in Fig. 2-2. In this system the food particles are frozen in a fluidized bed in which low-temperature air blows up through a conveyor chain, suspending the product being frozen. This air returns from the fluidized-bed conveyor to a heat exchanger that is the evaporator of a refrigeration unit. The refrigerating unit employs a reciprocating compressor and water-cooled condenser. A cooling tower, in turn, cools the condenser water rejecting heat to the atmosphere.
4. The design can be quantified by establishing certain values. Since the throughput of the plant has already been determined, the freezing capacity in lb/hr can be computed by deciding

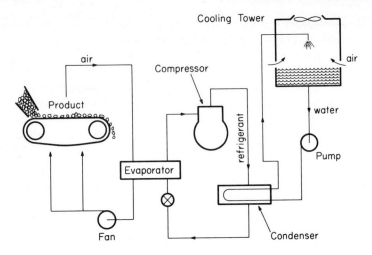

Fig. 2-2   Schematic flow diagram of freezing plant.

upon the number of shifts to be operated.   Assume that one
shift is selected so that now the refrigeration load can be calculated which, let us say, is 750,000 Btu/hr.   To proceed with
the design, the following parameters can be pinned down:

Temperature of chilled supply air	$- 20°F$
Temperature of return air	$- 10°F$
Refrigerant evaporating temperature	$- 30°F$
Condensing temperature	$110°F$
Temperature of condenser cooling water inlet	$85°F$
Temperature of condenser cooling water outlet	$95°F$

5. After the above values have been fixed, the individual components can be selected.   The flow of chilled air can be calculated
   to remove 750,000 Btu/hr with a temperature rise of 10°F.
   The conveyor length and speed must now be chosen to achieve
   the required rate of heat transfer.   The air-cooling evaporator
   can be selected from a catalog, because the air-flow rate, air
   temperatures, and refrigerant evaporating temperature fix the
   choice.   The compressor must develop 750,000 Btu/hr of
   refrigeration with an evaporating temperature of $-30°F$ and
   a condensing temperature of 110°F which is adequate information to select the compressor or perhaps a two-stage compression system.   The heat-rejection rate at the condenser exceeds
   the 750,000 Btu/hr refrigeration capacity by the amount of
   work added in the compressor and may be in the neighborhood

of 900,000 Btu/hr. The condenser and cooling tower can be sized based on the heat-flow rate and the water temperatures of 85°F and 95°F. Thus, a workable system can be designed.

Contrasting with the above procedures, an attempt to achieve an optimum system returns to the point where the first decisions are made. Such decisions as the location, size, and freezing concept should be considered in connection with one another rather than made independently. The choice of fluidized-bed freezing with a conventional refrigeration plant is only one of the concepts commercially available, to say nothing of the possibility (admittedly remote) of devising an entirely new concept. Other concepts are a freezing tunnel where the air blows over the top of the product, packaging the product first and immersing the package in cold brine until frozen, freezing the product with liquid nitrogen which is purchased in liquid form in bulk quantities. An example of the interconnection of decisions is that the best plant location using one concept may be different when using another. A compression refrigeration plant may be best located in the city as an extension of existing freezing facilities, and it may be unwise to locate it close to the producing area because of lack of trained operators. The liquid nitrogen freezing plant, on the other hand, is simple in operation and could be located close to the field and, furthermore, buttoned up for the idle off-season more conveniently than the compression plant. If the possibility of two or even three shifts were considered, the processing rate of the plant could be reduced by a factor of 2 or 3, respectively, for the same daily throughput.

Within the internal design of the compression refrigeration plant, the procedure was to select reasonable temperatures and then design each component around those temperatures and resulting flow rates. When approaching the design with the objective of optimization, all those interconnecting parameters would be left free to float, and the combination of values of these parameters found which results in the optimum—probably the economic optimum.

2-7 Preliminaries to the study of optimization An attempt to apply optimization theory to thermal systems at this stage is destined to frustration. There are a variety of optimization techniques available, some of which are studied in Chaps. 7 to 11. The first attempts to optimize a thermal system consisting of a dozen components, however, will encounter the task of predicting the performance of the system with given input conditions. This assignment is called *system simulation*. It must be studied first and will be considered in Chap. 5. For systems of any complexity, system simulation must be performed using a computer. When a computer is employed the performance characteristics making up the system could possibly be stored as tables, but a far more efficient and useful form is equation-type formulation. Translating catalog tables into equations, called *component simulation* or *mathematical modeling* is a routine preliminary to system simulation and will be treated in Chap. 4.

Finally, since optimization presupposes a criterion, which in engineering practice is often an economic one, a review of investment economics in Chap. 3 would be appropriate.

The sequence to be followed in the ensuing studies, then, will be (1) economics, (2) mathematical modeling, (3) system simulation, and (4) optimization.

PROBLEMS

2-1. Location S in Fig. 2-3 is an adequate source of water, and locations A, B, and C, which are all at the same elevation as S, are points at which water must be provided at the following rates of flow:

A: 40 gpm B: 55 gpm C: 25 gpm

The demands for water at A and C occur intermittently and only during the working day, although they may at certain times be coincident. The

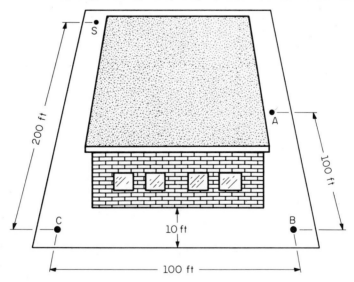

Fig. 2-3 Supply and consumption points in water-distribution system.

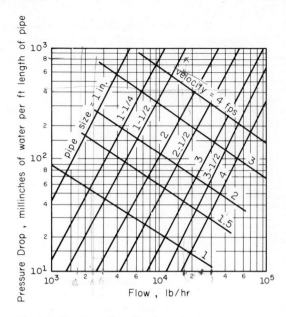

Fig. 2-4 Pressure drop in pipe.

demand for water at B occurs only at times during nonworking hours, and also intermittently.

Ground-level access exists in a 10-ft wide border which surrounds the building. Access is not permitted over, through, or under the building.

(a) Describe all the concepts of workable methods that you can devise to achieve the assignment.

(b) The influence of such factors as the expected life of the system has resulted in the decision to use a system in which a pump delivers water into an elevated storage tank which supplies the piping system. A water-level switch starts and stops the pump.

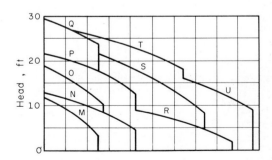

Fig. 2-5 Pump performance curves.

Design the system; i.e., specify the piping network, all pipe sizes using pressure-drop data from Fig. 2-4, the model of the pump chosen from Fig. 2-5, and the elevation of the storage tank. (Neglect the pressure drop in the pipe fittings and pressure conversions due to kinetic energy.)

Pipe section	Pipe size, in.	Design flow, gpm		Δp, ft of water	
		Day	Night	Day	Night
(S to A, for example)					
A+					

Pump _____   Storage tank elevation _____ ft

(c) Review the design and list the decisions that precluded a possible optimization later in the design.

2-2. A heating and ventilating system for a public, indoor swimming pool is to be designed.

Specifications:
    Pool water temperature   78°F
    Indoor air dry-bulb temperature   80°F
    Outdoor design temperature   10°F
    Outdoor design relative humidity   80%
    For odor control, the minimum rate of outdoor air for
        ventilation is 45,000 lb/hr
    When air is introduced into the space, its temperature
        should be between 100°F and 140°F for comfort
        reasons
Construction features:
    Pool dimensions   50 × 150 ft
    Essentially the structure has masonry walls and a glass
        roof
        Glass area = 20,000 ft²
        Wall area = 7,200 ft²
        U value of wall = 0.21 Btu/(hr)(ft²)(°F)
Supplementary information:
    Rate of evaporation of water from the pool in lb/hr is

$$(0.2)(\text{area, ft}^2)(p_w - p_a)$$

where p_w = water vapor pressure at pool temperature,
           psia
      p_a = partial pressure of water vapor in surround-
           ing air, psia

A choice of the type of window must be made—single, double, or triple, having the following heat-transfer coefficients in Btu/(hr)(ft²)(°F):

	Single	Double	Triple
Outside air film	6.0	6.0	6.0
Glass (between external and internal surfaces)	20.8	1.42	0.60
Inside film	2.0	2.0	2.0

The basic assignment of the system is:

To maintain the indoor temperature when design conditions prevail outdoors
To abide by the constraints
To prevent condensation of water vapor on the inside of the glass

(a) Describe at least two different concepts of accomplishing the objectives of the design. Use schematic diagrams if useful.

(b) Assume that the concept chosen is one where outdoor air is drawn in, heated, introduced to the space, and an equal quantity of moist room air is exhausted. Should additional heating be required, it is provided either by heating recirculated air or by radiators or convectors around the perimeter of the building.

Perform the design calculations in order to specify the following:

Flow rate of ventilation air, lb/hr _____
Temperature of air entering space, °F _____
Type of glass selected _____
Temperature of inside surface of glass, °F _____
Condition of indoor air: dew point, °F _____
Rate of water evaporated from pool, lb/hr _____
Heat loss by conduction through walls and glass, Btu/hr _____
Heat supplied to raise temperature of ventilation air from outdoor temperature to room temperature, Btu/hr _____
Heat supplied to raise temperature of ventilation air from room temperature to supply temperature, Btu/hr _____
Heat to recirculated air or perimeter radiation, Btu/hr _____

(c) Review the design and list the decisions that precluded a possible optimization later in the design.

2-3. You have just purchased a remote, uninhabited island where you plan to give parties for the entertainment of your friends. These parties are to be gala events, sometimes lasting until as late as 10:30 P.M., so you wish to install an electric power system that will provide at least 8 kW of lighting.

(a) List and describe, in several sentences, three methods of generating power in this remote location, assuming that needed equipment and supplies can be transported to the island.

(b) Assume that the decision has been made to produce the electric power by means of an engine-driven generator. Specifically, an engine will be direct-connected to a generator that delivers power to bulbs. Choices that are available are as follows:

Bulbs: 100-W bulbs at 115 V. The bulbs may be connected in parallel, in series, or in combination, and the current flowing through a bulb must be above 0.7(rated current) to obtain satisfactory lighting efficiency, but below 1.1(rated current) to achieve long life.

Engines: A choice can be made between two engines whose power deliveries at wide-open throttle are: → 8KW   → 35Hz

	800 rpm	1,200 rpm	1,600 rpm	2,000 rpm	2,400 rpm	2,800 rpm
Engine 1, hp	4.8	7.0	10.9	13.6	15.0	16.6
Engine 2, hp	5.8	9.5	13.2	16.5	18.5	20.1

Generators: Single phase, alternating current. There is no adjustment of output voltage possible except by varying the speed. The frequency must be greater than 35 Hz in order to prevent flickering of the lights. Neglect the voltage drop and power loss in the transmission lines.

A choice can be made between the following two generators:

	Poles	Efficiency	Maximum allowable current, A	Output voltage at given speed, rpm				
				500	1,000	1,500	2,000	2,500
Generator 1	2	0.72	100	31	60	89	118	146
Generator 2	4	0.83	50	50	98	146	185	230

Make the following design selections and specifications:

Engine number ___1___
Generator number ___2___
Engine-generator speed ___2400___ rpm
Frequency ___40___ Hz
Engine-power output ___15___ hp = ___11.2___ kW
Generator-power output ___11.5___ kW
Voltage ___230___ V
Current ___50___ A
Number of bulbs and circuiting:

(c) Review the design and list the decisions that precluded a possible optimization later in the design.

3
Economics

3-1 Introduction The basis of most engineering decisions is economic. Designing and building a device or system that functions properly is only a part of the engineer's task. The device or system must, in addition, be economical—it must show an adequate return on the investment. In the study of thermal systems, one of the key ingredients is optimization, and the function that is most frequently optimized is the potential profit. Often, we seek the design having minimum first cost or, more frequently, the minimum total owning and operating cost during the life of the facility.

It is unpopular in some circles to suggest that decisions are made purely on the basis of monetary considerations. The fact is that many noneconomic factors affect the decisions of industrial organizations. Decisions are frequently influenced by political concerns, such as zoning regulations, or by social concerns, such as the displacement of workers, or by air or stream pollution. Aesthetics also have their influence, as for example, when extra money is spent in order to make a new factory building attractive.   These social or aesthetic concerns almost always require the outlay of extra money, so they once again revert to economic questions, such as how much a firm is willing or able to spend for such decisions as locating a plant where the employees will live in a district with good schools.

This chapter first discusses the consequence of interest charges on both lump sums and series of payments.   This study leads to formulas and factors for ease of calculation.   Presented next are several most-used methods of making economic evaluations and comparisons.   The final topic considered in this chapter is taxes and several of the most-used methods of computing depreciation— a quantity used in tax calculations.

3-2 Interest   Interest is the rental charge for the use of money. When renting a house, a tenant pays rent but also returns the use of the house back to the owner after the stipulated period.   In its simplest form, the borrower of money pays interest for the use of the money.   Complicating the interest calculation is the practice that the borrower of the money pays the interest at stated periods through the term of the loan—for example, every 6 months or every year.   Other standard interest calculations are those that surround series of payments, called *annuities*.   What is the annual payment, for example, that will retire a certain loan in 10 years, while interest is being charged on the unpaid balance?

The seven frequently encountered interest-paying situations that will be explained in the next sections are:

1. Simple interest
2. Compound interest
3. Present worth
4. Sinking fund
5. Series compound amount
6. Series present worth
7. Capital recovery

3-3 Simple interest   Harry Smith borrows $500 from his grand-mother to be repaid in 3 years.   He agrees to pay her 5 percent interest per year.   At the end of 3 years, Harry brings to his grandmother

$$\$500 + (\$500)(0.05)(3 \text{ years}) = \$575$$

The $500 is called the *principal*, and the *rate of interest* is the interest earned per unit principal per unit time.   As Harry has made the calculation, the interest is called *simple interest*. The entire amount S to be repayed at the future time is

$$S = P + Pni = P(1 + ni)$$

where P = principal
i = rate of interest per period
n = number of periods

3-4 Compound interest "Dear boy," said Harry's grandmother, "You have not read our contract carefully. The contract states that you will pay 5 percent interest per year, *compounded annually*. At the end of the first year, therefore, you owed me

$$\$500 + (\$500)(0.05) = \$525$$

At the end of the second year, you owed me

$$\$525 + (\$525)(0.05) = \$551.25$$

Now, at the end of the third year, you owe me

$$\$551.25 + (\$551.25)(0.05) = \$578.81"$$

The following is the pattern that develops in computing compound interest on principal P with interest i per period:

Period	Interest during period	Amount S at end of period
1	$P(i)$	$P + Pi = P(1 + i)$
2	$P(1 + i)i$	$P(1 + i) + P(1 + i)i = P(1 + i)^2$
. . .	. . . .	. . . .
n	$P(1 + i)^{n-1}i$	$P(1 + i)^{n-1} + P(1 + i)^{n-1}i = P(1 + i)^n$

In computing compound interest, then,

$$S = P(1 + i)^n = P(\text{CAF}) \tag{3-1}$$

where CAF is called the *compound-amount factor* and is equal to $(1 + i)^n$. Values of CAF, as well as other interest and annuity factors are tabulated in most textbooks in engineering economics. When a digital computer is available, one may compute his own set of tables as suggested in Prob. 3-1.

Example 3-1 You invest \$3,000 in a credit union which pays 5 percent per year, compounded semiannually. What is the total sum that has accumulated at the end of 5 years?

Solution Interest rates, by convention, are expressed on an annual basis, even
though the compounding period is different from 1 year.   The specification of 5 percent per year, compounded semiannually, means that the
interest rate is 2.5 percent per half-year and the period is a half-year.
The interest rate of 5 percent in this situation is sometimes called the
nominal rate of interest, since the annual rate of return is not exactly 5
percent per year, but somewhat greater.   Applying Eq. (3-1) to the
problem at hand, the principal P = \$3,000, the rate of interest i =
0.025 per period, and the number of periods n = 10.

$$S = (\$3,000)(1 + 0.025)^{10} = (\$3,000)(\text{CAF-2.5\%-10})$$

From interest tables, CAF-2.5%-10 = 1.2801.   Then

$$S = (\$3,000)(1.2801) = \$3,840.30$$

3-5 Present worth   A companion calculation to that of compound
interest is the present-worth calculation.   *Present worth* is the value
of a sum of money at the present time that, with compounded interest, will have a specified value at a certain time in the future.   In
Eq. (3-1), therefore, S is known and P is to be determined.

$$P = \frac{S}{(1 + i)^n} = (\text{PWF})S \tag{3-2}$$

The abbreviation PWF designates the *single-payment present-worth
factor* and is simply the reciprocal of the CAF for given values of i
and n.

Example 3-2   A father wishes to invest a sum of money when his son begins
elementary school such that the accumulated amount will be \$5,000
when the son begins college 12 years later.   The father is able to invest
the money where it will draw interest at the rate of 4 percent compounded
semiannually.   What amount must be invested?

Solution The value that must be calculated is the present worth of \$5,000,
earning 2 percent interest per period, for a time 24 interest periods
hence.   From interest tables,

PWF-2%-24 = 0.6217

The present worth, then, is

$$P = (\$5,000)(0.6217) = \$3,108.50$$

3-6 Series compound amount   In many business and personal
financial situations, equal payments are made at regular intervals
in order to accumulate an amount some time in the future.   The
sum earns interest on the progressively increasing amount.   A series
of equal payments at regular periods of time is called an *annuity.*
Such a series might occur in a personal investment program where
equal payments are set aside from an income.

Suppose that \$100 is invested at the end of every year and draws 6 percent interest on the accumulated amount, compounded annually. At the end of the first period the accumulated sum S is only the \$100 payment which has just been made.

At the end of two years,

$$S = (\$100)(1 + i) + \$100 = \$100[(1 + i) + 1]$$

At the end of n years,

$$S = (\$100)[(1 + i)^{n-1} + (1 + i)^{n-2} + \cdots + (1 + i) + 1]$$

The series in the brackets,

$$\sum_{j=0}^{n-1} (1 + i)^j$$

can be represented by the expression

$$\sum_{j=0}^{n-1} (1 + i)^j = \frac{(1 + i)^n - 1}{i} \qquad (3\text{-}3)$$

Then,

$$S = R\left[\frac{(1 + i)^n - 1}{i}\right] = R(\text{SCAF})$$

where R = regular payment

SCAF = series compound-amount factor

That the expression in Eq. (3-3) correctly represents the series can be shown by mathematical induction. The steps in mathematical induction are codified as follows:

1. Prove that the expression is true when the index n equals 1.
2. Assuming that the statement is true for the index n, prove that it is true for index $n + 1$.

Applying these two steps of mathematical induction to test the truth of Eq. (3-3),

Step 1   When $n = 1$,

$$\sum_{0}^{0} (1 + i)^j = (1 + i)^0 = 1 : \frac{(1 + i)^1 - 1}{i} = \frac{i}{i} = 1$$

so the expression is true for $n = 1$.

Step 2   Assuming that Eq. (3-3) is correct, we must prove that

$$\sum_{j=0}^{(n+1)-1} (1+i)^j = \frac{(1+i)^{n+1} - 1}{i}$$

To prove this, add $(1+i)^n$ to both sides of Eq. (3-3).

$$\sum_{j=0}^{n-1} (1+i)^j + (1+i)^n = \frac{(1+i)^n - 1}{i} + (1+i)^n$$

$$\sum_{j=0}^{n} (1+i)^j = \frac{(1+i)^n - 1 + i(1+i)^n}{i}$$

$$= \frac{(1+i)^n(1+i) - 1}{i}$$

Thus,

$$\sum_{j=0}^{(n+1)-1} (1+i)^j = \frac{(1+i)^{n+1} - 1}{i}$$

which is the proof required by step 2.

Example 3-3   A man at the age of 60, in order to plan for the future, decides to invest \$1,000 at the end of the present year, and an equal amount every year thereafter.   The money is invested in bonds that draw 6 percent, compounded annually.   What will be the accumulated amount when he reaches the age of 80?

Solution   The accumulated amount S is

$$S = (\text{SCAF})(\$1,000)$$

where

$$\text{SCAF-6\%-20} = 36.7856$$

At the age of 80 the man, therefore, has \$36,785.60 with which to enjoy life.

It should be emphasized that the series compound-amount factor is based on the first payment being made at the *end* of the first year and at the *end* of each succeeding year.   In a similar manner, the next three factors to be studied—the sinking fund, series present-worth factor, and capital-recovery factor—are also computed on the basis of payments or withdrawals at the *ends* of the years.

3-7   Sinking fund   The companion calculation to the series compound-amount factor is the sinking-fund factor.   If the amount that is to be accumulated by some specific time in the future has been specified, the sinking-fund calculation determines the regular payments that must be made in order to accumulate the amount. In equation form

$$R = \frac{S}{\text{SCAF}} = S\left[\frac{i}{(1+i)^n - 1}\right] = S(\text{SFF}) \qquad (3\text{-}4)$$

where

SFF = sinking-fund factor

Example 3-4 A new machine has just been installed in a factory and the management wants to set aside and invest equal amounts each year starting 1 year from now so that \$16,000 will be available in 8 years for the replacement of the machine. The interest received is 6 percent, compounded annually. How much should be set aside each year?

Solution The annual payment R is

$$R = S(\text{SFF-}6\%\text{-}8) = (\$16,000)(0.101)$$
$$R = \$1,616 \text{ per year}$$

3-8 Series present worth In the situation where the series compound amount and sinking fund were applicable, regular payments were made in order to accumulate a specified amount at the end of a certain period of time. The reverse procedure is where an amount exists at the start and equal withdrawals are made periodically so that the amount is exactly used up after a specified time. Interest is earned on the amount that has not yet been withdrawn. Let P be the original amount, with an amount R being withdrawn at the end of the first and subsequent periods. The amount P earns interest at the rate of i percent during the first period.

After one period, the current value is

$$P(1+i) - R$$

After two periods, the current value is

$$[P(1+i) - R](1+i) - R = P(1+i)^2 - R[(1+i) + 1]$$

If the amount is to be exhausted at the end of n periods,

$$P(1+i)^n - R[(1+i)^{n-1} + \cdots + 1] = 0$$

The series in the brackets has already been evaluated and is expressed in Eq. (3-3). Therefore,

$$P = R\frac{(1+i)^n - 1}{i(1+i)^n} = R(\text{SPWF}) \qquad (3\text{-}5)$$

where

$$\text{SPWF} = \text{series present-worth factor}$$

Example 3-5 A finance company advertises that a borrower pays $6.62 interest during the course of a year if he borrows $100 and repays it in 12 equal monthly payments. What is the nominal rate of interest charged if the interest is compounded monthly?

Solution The series present-worth calculation is applicable in a negative sense because the borrower begins with a negative sum of $100 on which he receives negative interest payments, and mean-while he makes negative withdrawals.

During the year, the borrower pays $100 plus the interest of $6.62, so his equal monthly payments are $106.62/12 = $8.89. Applying Eq. (3-5),

$$\$100 = (\$8.89)(\text{SPWF})$$

so

$$\text{SPWF} = 11.25$$

Referring to the table of the SPWF for 12 periods, we find that the interest is 1 percent per period.

The nominal rate of interest is, therefore, 12 percent per year.

3-9 Capital-recovery factor The companion factor to the SPWF is the capital-recovery factor CRF which when multiplied by the original amount P gives the regular withdrawal amount that exhausts the money after a specified time.

$$R = \frac{P}{\text{SPWF}} = P\,\frac{i(1 + i)^n}{(1 + i)^n - 1} = P(\text{CRF}) \qquad (3\text{-}6)$$

A summary of the interest and annuity formulas discussed so far is presented in Table 3-1.

Several patterns emerge from Table 3-1. The factors consist of three reciprocal pairs—CAF and PWF, SCAF and SFF, and SPWF and CRF. If a table of the CAF is available, for example, the PWF table is unnecessary, because present-worth calculations could be made by dividing the future amount by the CAF.

A relation exists between the SCAF and the SPWF that is comparable to the one that exists between the SFF and the CRF. The relation is the translation of the worth between the present and future time. All the factors pertain to the values of a series of payments or withdrawals. The SCAF establishes the future worth, while the SPWF establishes the present worth, so SPWF = (SCAF)(PWF).

Table 3-1   Summary of interest and annuity formulas

Symbol	Name	Equation	Situation
CAF	Compound amount	$(1 + i)^n$	Future amount $= $ (CAF)(present amount)
PWF	Present worth	$\dfrac{1}{(1 + i)^n}$	Present amount $= $ (PWF)(future amount)
SCAF	Series compound amount	$\dfrac{(1 + i)^n - 1}{i}$	Future amount $= $ (SCAF)(regular payment)
SFF	Sinking fund	$\dfrac{i}{(1 + i)^n - 1}$	Regular payment $= $ (SFF)(future amount)
SPWF	Series present worth	$\dfrac{(1 + i)^n - 1}{i(1 + i)^n}$	Present amount $= $ (SPWF)(regular payment)
CRF	Capital recovery	$\dfrac{i(1 + i)^n}{(1 + i)^n - 1}$	Regular payment $= $ (CRF)(present amount)

3-10   Uniform-gradient series   The factors, SCAF, SFF, SPWF, and CRF are all applicable where the payments or withdrawals are *uniform*.   The situation may arise where the present worth of a series of *increasing* payments is sought.   Probably, the most common example of this need is the determination of the cost of maintenance on equipment which is expected to progressively increase as the equipment ages.

The gradient present-worth factor GPWF applies to the case where there is no cost during the first year, a cost G at the end of the second year, $2G$ at the end of the third year, and so on, as shown graphically in Fig. 3-1.   The straight line in Fig. 3-1 starts at the end of year 1 in order to represent the typical maintenance situation where there is no maintenance required during the first year, an amount G at the end of the second year, and $2G$ at the end of the third year.

The present worth of this series of increasing payments is the sum of the individual present worths.

$$\text{Present worth} = \frac{G}{(1 + i)^2} + \frac{2G}{(1 + i)^3} + \cdots + \frac{(n - 1)G}{(1 + i)^n}$$

This series can be calculated by the formula

$$\text{Present worth} = G \left\{ \frac{1}{i} \left[\frac{(1 + i)^n - 1}{i(1 + i)^n} - \frac{n}{(1 + i)^n} \right] \right\}$$
$$= G(\text{GPWF}) \tag{3-7}$$

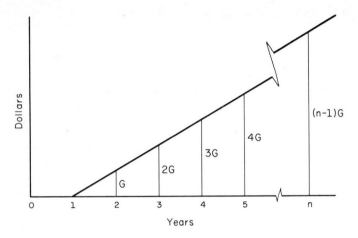

Fig. 3-1 Uniform-gradient series of payments.

3-11 Variations in use of basic factors Many economic calculations consist of nothing more than a proper decision as to which
of the previously mentioned factors is applicable, then a multiplication by that factor. The tables of interest and annuity factors
can also be used as building blocks to solve problems that are variations of the basic cases. Several examples of such variations are
discussed in Secs. 3-12 to 3-14, for instance, changes in the midst
of series payments, shifts of payment time, and different periods
of payment and compounding.

3-12 Change in midstream When a plan is to be established for
a series of payments (or withdrawals), the amounts of the payments
are based on a specified interest rate, accumulated amount, and
number of periods. It is not uncommon for a program which starts
on one basis to be altered at some later date to accommodate changes
in conditions. This class of problems can, in general, be solved by
establishing the value at the point of change and making a new calculation for the remaining time.

> **Example 3-6** A sinking fund is established such that $12,000 will be available
> to replace a facility at the end of 10 years. At the end of 4 years,
> following the fourth uniform payment, management decides to retire
> the facility at the end of 9 years of life. Using an interest rate of 6
> percent, compounded annually, what are the payments during the first
> 4 and during the last 5 years?

Solution   On the original plan, the payments that were to be made for 10 years
were

($12,000)(SFF-6%-10) = (12,000)(0.0759) = $910.80

The accumulated value at the end of 4 years is

($910.80)(SCAF-6%-4) = (910.80)(4.3746) = $3,984.38

This accumulated value will draw interest for the next 5 years, but the
remainder of the $12,000 must be provided by the five additional pay-
ments plus interest.

R(SCAF-6%-5) = $12,000 − (3,984)(CAF-6%-5)

so the payments during the last 5 years must be

$$R = \frac{12,000 - (3,984)(1.3382)}{5.6370} = \$1,183$$

3-13   Shifts in time of payment   The tables of factors, such as those
involving a uniform series of payments, apply to payments starting
at a specified time.   The SCAF, for example, assumes that the first
payment occurs at the end of the first period.   Suppose instead that
the first payment is made at time zero, and one period elapses after
the final payment.   In place of the series listed in Eq. (3-3), the
accumulated value of the series at the end of n years would be

$$(1 + i)^n + (1 + i)^{n-1} + \cdots + (1 + i)$$

This series is identical to Eq. (3-3) if Eq. (3-3) is multiplied by
$(1 + i)$.   Thus, the series compound-amount factor if the payments
start at the beginning of the first period instead of at the end would
be (SCAF)$(1 + i)$.

3-14   Different payment and compounding periods   The tables of
factors for uniform payments, SCAF, and SFF, for example, are
based on the assumption that the payment period and the compounding periods are identical.   A situation may arise where these
periods are different.   For example, assume that uniform annual
investments of R are made, beginning at the end of the first year.
The interest, however, is compounded semiannually and is at a rate
of i per half-year.   What is the accumulated amount at the end of
N years?

The accumulated amount builds as follows:

End of, years	Accumulated amount
½	0
1	R
1½	$R(1 + i)$
2	$R[(1 + i)^2 + 1]$
2½	$R[(1 + i)^3 + (1 + i)]$
3	$R[(1 + i)^4 + (1 + i)^2 + 1]$
N	$R[(1 + i)^{2(N-1)} + (1 + i)^{2(N-2)} + \cdots + 1]$

The series within the bracket after N years can be written

$$[(1 + i)^2]^{(N-1)} + [(1 + i)^2]^{(N-2)} + \cdots + 1$$

which is similar to the series in Eq. (3-3) except that the semiannual interest rate of i must be converted to an annual rate i_a by the equation

$$(1 + i)^2 = 1 + i_a$$

Example 3-7 The sum of $100 is invested at the end of the first and each succeeding year for 5 years while interest at a 6 percent annual rate is compounded semiannually. What is the accumulated amount at the end of 5 years?

Solution The comparable annual rate i_a is

$$i_a = (1 + i)^2 - 1 = (1 + 0.03)^2 - 1 = 0.0609$$

The accumulated amount at the end of 5 years is

$$S = (100)(SCAF\text{-}6.09\%\text{-}5) = (100)(5.6472) = \$564.72$$

It should be noted that the translation of the SCAF tables is not applicable for noninteger values of N.

3-15 Investment economics One of the most important functions of economic techniques in engineering enterprises is to evaluate proposed investments. If an organization has money which it wishes to invest, an economic study can indicate which is the most favorable choice among several alternates. An investment analysis may indicate whether or not a project should be undertaken at all. Often an organization must borrow money—by selling stock in the case of a private enterprise, or by issuance of bonds by a public organization—in order to carry out the project. The investment analysis seeks to predict whether a profit can be shown even after the principal and interest are repaid. Although future costs and prices cannot be predicted with certainty, an investment analysis indicates the likelihood of profitability.

Four different methods of investment analysis will be explained:

1. Present worth
2. Uniform annual cost
3. Rate of return
4. Break-even point

The methods will be illustrated by applying them to an example, which while simplified does include the basic quantities that usually influence an investment, such as,

1. First cost
2. Salvage value
3. Operating expense
4. Income

All of these quantities are affected by appropriate interest charges. Some of the complications usually included in more elaborate investment analyses, such as, depreciation, tax, and inflation considerations, will be omitted temporarily so that the basic procedures used in these methods can be emphasized.

Example 3-8 An investor is evaluating the attractiveness of constructing and operating an office building for rental purposes. Two types of construction are being evaluated, one a more expensive one with longer life than the other. The potential rental income will be the same on the two buildings, but operating costs and salvage values differ. Table 3-2 shows the economic data.

3-16 Present-worth method In this technique, all of the costs and incomes are translated into present worths. In order to determine

Table 3-2   **Economic data on buildings in Example 3-8**

	Building A	Building B
Useful life	20 years	30 years
First cost	$110,000	$140,000
Annual operating and maintenance cost	$ 11,000	$ 9,000
Annual income from rent	$ 20,300	$ 20,300
Salvage value	$ 10,000	$ 20,000

the present worth of a future sum, the interest rate must be established, and let us assume that 6 percent per year is applicable. The comparison of the attractiveness of the two investments in Example 3-8 is made slightly complicated by the fact that the expected lives of the two buildings are different. This difficulty can be surmounted, however, by renewing the building until the two lives come out even. That period would be 60 years, in this case, requiring two renewals of building A and one renewal of building B.

Let us concentrate first on building A. The present worth of the first cost is \$110,000, because that sum is applicable at the present time.

At the end of 20 years, the building has a salvage value of \$10,000 which may be deducted from the \$110,000 first cost to rebuild the structure at that time. The present worth of the rebuilding cost 20 years from now is

$$(110,000 - 10,000)(\text{PWF-6\%-20})$$
$$= (100,000)(0.3118) = \$31,180$$

One more renewal of the building is needed in order to attain the 60-year life, and this renewal occurs at 40 years, which represents a present worth of $(100,000)(\text{PWF-6\%-40})$ or

$$(100,000)(0.0972) = \$9,720$$

Finally, at the end of 60 years, the salvage value of \$10,000 has a present worth of $(10,000)(0.0303) = \$303$ which is an income and, thus, deductible from the present-worth costs.

The operating expenses are \$11,000 per year and the present worth of these expenditures for a 60-year period is

$$(11,000)(\text{SPWF-6\%-60}) = (11,000)(16.161) = \$177,778$$

The present-worth values applicable to building B are computed in a corresponding manner, although only one renewal of the building is required to attain the 60-year total life. A summary of the present-worth values of the two buildings is shown in Table 3-3.

The present-worth costs of building B are less than those of building A, and the expected income from the buildings is identical, so building B is the preferred choice of investment.

Present-worth calculations are in widespread use, although they do pose an added difficulty if the lives of two possible investments are different. The renewal assumptions must be realistic in order for the analysis to be valid. Suppose, for example, that the

Table 3-3   Present-worth values of costs of buildings A and B

	Building A	Building B
First cost	$110,000	$140,000
First renewal	31,180	20,890
Second renewal	9,720	
Salvage value	−303	−606
Operating and maintenance	177,778	145,454
	$328,375	$305,738

need for this building beyond 30 years is questionable.   The first renewal of building A, then, that extends the life beyond 30 years is invalid.

3-17 Annual-cost method   In contrast to the present-worth method where all annual costs were translated to a present-worth cost, the annual-cost method translates all nonannual costs to an annual cost basis.   The translation requires knowledge of the interest rate, which again will be assumed to be 6 percent.

In building A, for example, the nonannual costs are the first cost and the salvage value.   The $110,000 first cost expressed as an annual cost over the 20-year life of the building is

$$(110,000)(\text{CRF-6\%-20}) = (110,000)(0.0872) = \$9,590$$

The salvage value of $10,000 20 years hence is equivalent to annual amounts of $(10,000)(\text{SFF-6\%-20}) = \$270$ which is a negative cost or a return.   The summary of annual costs for the two buildings is shown in Table 3-4.

Since building B has a lower annual cost than building A, it is the more attractive investment—a conclusion identical to that indicated by the present-worth method.   The annual-cost method has two advantages over the present-worth method.   One is that there

Table 3-4   Annual costs of buildings A and B

	Building A	Building B
First cost	$ 9,590	$10,170
Operating and maintenance	11,000	9,000
Salvage value	−270	−250
Total	$20,320	$18,920

is no complication introduced when the two prospective investments have different lives. The other is that it is more natural for most people to think in terms of an annual cost than in terms of a present worth.

3-18   Rate-of-return method   In both the present-worth method and the annual-cost method, a rate of interest of 6 percent was assumed. In the rate-of-return method, no assumption is made regarding the interest rate. Instead, the money for the investment is considered to be in hand, and the rate of return is calculated as though it were an interest rate received by making an outside investment.

Concentrating on building A, the statement of the economic situation is that the investment expressed as an annual cost at an unknown rate of return yields the net income. The net income is the annual rental minus the annual operating expenses. Thus,

$$(110,000)(\text{CRF-?\%-20}) - (10,000)(\text{SFF-?\%-20}) = 20,300 \\ - 11,000$$

A trial-and-error process can be used to determine the unknown rate of return i. For example, first try $i = 8$ percent.

$$(110,000)(0.1019) - (10,000)(0.0219) = 10,990 \neq 9,300$$

Next try $i = 6$ percent.

$$(110,000)(0.0872) - (10,000)(0.0272) = 9,320 \approx 9,300$$

The rate of return on the \$110,000 investment is therefore, 6 percent.

By a similar procedure, the rate of return on building B can also be found. With a trial of an 8 percent return the expected net income is \$12,144, which is larger than the actual income of \$11,300. A trial of 6 percent shows a return of \$9,912, so by interpolation, the rate of return is 7.24 percent.

As in the previous two methods, the rate-of-return method shows building B to be a more profitable investment than building A.

3-19   Break-even point   This method assumes that money is borrowed at a specified rate of interest and that the loan is paid off as rapidly as possible with no profits extracted. The break-even point is defined as the time where the loan is paid off and profits begin flowing to the investor. In contrast to the rate-of-return method where the interest rate was unknown, now the life is unknown. An interest rate of 6 percent will again be assumed. We seek a value of n such that the following equation is satisfied for building A:

$$(110,000)(\text{CRF-}6\%\text{-}n) - (10,000)(\text{SFF-}6\%\text{-}n) = 9,300$$

The salvage value of $10,000 is assumed to be independent of the age, which may or may not be 20 years at the break-even point. Try $n = 18$ years.

$$(110,000)(0.0924) - (10,000)(0.0324) = 9,840 \neq 9,300$$

Try 20 years.

$$(110,000)(0.0872) - (10,000)(0.0272) = 9,320 \approx 9,300$$

The break-even point for building A is 20 years, so after this period of time the income has paid off the investment, but the life of the building has also expired so no profits are available to the investor.

In contrast, building B has a break-even point of $21\frac{1}{2}$ years, and from that point until the end of the useful life of 30 years, the investor will pocket profit from the operation of the building.

3-20 Taxes The money for operating the government and for financing services provided by the government derives primarily from taxes. The inclusion of taxes in an economic analysis is often important because in some cases taxes may be the deciding factor of whether or not to undertake the project. In certain other cases, the introduction of tax considerations may influence which of two alternates will be the most economically attractive.

In most sections of the United States, property taxes are levied by a substate taxing district in order to pay for schools, city government, and services, and perhaps park and sewage systems. Theoretically, the real estate tax should decrease as the facility depreciates, resulting in lower real estate taxes as the facility ages. Often on investments such as buildings the tax, as a dollar figure, never decreases. It is a common practice, therefore, to plan for a constant real estate tax when making the investment analysis. The effect of the tax is to penalize a facility which has a high taxable value.

Federal corporation income taxes on any but the smallest enterprises run in the neighborhood of 50 percent of the profit before taxes. Income tax is usually a much more significant factor in the economic analysis than is property tax, so income tax will be discussed further. An ingredient of income tax calculations is *depreciation* which will be discussed in the next section.

3-21 Depreciation Depreciation is an amount that is listed as an annual expense in the tax calculation to allow for replacement of the facility at the end of its life. There are numerous methods of computing depreciation which are permitted by the Internal Revenue Service, such as straight line, sum-of-the-year's digits, and double-rate declining balance. The first two methods will be explained in this section.

Straight-line depreciation simply consists of dividing the difference of the first cost and salvage value of the facility by the number of years of tax life. The result is the annual depreciation. The tax life to be used is prescribed by the Internal Revenue Service and may or may not be the same as the economic life used in the economic analysis.

In the sum-of-the-year's digits or SYD method, the depreciation for a given year is represented by the formula

$$\text{Depreciation (\$)} = \frac{2(N - t + 1)}{N(N + 1)} (P - S) \tag{3-8}$$

where N = tax life, years
$\quad\quad t$ = year in question
$\quad\quad P$ = first cost, \$
$\quad\quad S$ = salvage value, \$
If the tax life is 10 years, for example, the depreciation is as follows:

$\quad\quad \frac{10}{55}(P - S)$ for the first year

$\quad\quad \frac{9}{55}(P - S)$ for the second year

$\quad\quad . \ . \ . \ . \ . \ . \ . \ . \ . \ . \ . \ . \ . \ . \ . \ . \ . \ . \ . \ .$

$\quad\quad \frac{1}{55}(P - S)$ for the tenth year

A comparison of the depreciation rates calculated by the straight-line method with that of the SYD method shows that the use of the SYD method permits greater depreciation in the early portion of the life. When using the SYD method, then, the income tax that must be paid early in the life of the facility is less than by the straight-line method, although near the end of the life the SYD tax is greater than straight line. The total tax paid over the tax life of the facility is the same by either method, but the advantage of using the SYD method is that more of the tax is paid in later years which is advantageous when considering the time value of money.

A situation where the straight-line method has an advantage is where there is the prospect of an increase in the tax rate. If the rate jumps, it is better to have paid the low tax on a larger fraction of the investment.

3-22 Influence of income tax on economic analysis To show the effect of depreciation and federal income tax, consider the following simple example of choosing between alternate investments A and B for which the following data apply:

	Alternate A	Alternate B
First cost	$200,000	$270,000
Life	20 years	30 years
Salvage value	0	0
Annual income	$ 50,000	$ 50,000
Out-of-pocket expense	$ 14,600	$ 8,900
Real estate taxes and insurance (5%)	$ 10,000	$ 13,500

A calculation of the annual cost of alternate A without inclusion of the income tax is:

First cost on annual basis ($200,000)(CRF-6%-20)

$$= \$17,440$$

Out-of-pocket expenses   14,600

Real estate tax and insurance   10,000

Total $42,040

Similar calculations for alternate B are:

First cost on annual basis ($270,000)(CRF-6%-30)

$$= \$19,600$$

Out-of-pocket expense   8,900

Real estate tax and insurance   13,500

Total $42,000

The economic analysis of alternates A and B shows approximately the same annual costs and incomes—and, in fact, the example was rigged to accomplish this.

In the computation of profit on which to pay income tax, the actual interest paid is listed as an expense, and if straight-line depreciation is applied, the expenses for the first year of the two alternates are:

First-year expenses	Alternate A	Alternate B
Depreciation	$10,000	$ 9,000
Interest (6% of first cost)	12,000	16,200
Out-of-pocket, tax and insurance	24,600	22,400
Total expenses	$46,600	$47,600
Profit = income − expenses	3,400	2,400
Income tax (50% of profit)	1,700	1,200

So a higher income tax must be paid on alternate A than on alternate B during the early years. In the later years of the project, a higher tax will be paid on alternate B. The example shows that even though the investment analysis indicated equal profit on the two alternatives, the inclusion of income tax shifts the preference to alternate B. The advantage of alternate B is that the present worth of the tax payments is less than in alternate A.

3-23 Summary There are many levels of economic analysis higher than that achieved in this chapter. The complications encountered in accounting, financing, and tax computations involve sophistications beyond those presented here. Achievement of the second level of economic analysis could be considered the accomplishment of this chapter. The first level would be a trivial one of simply totaling costs with no consideration of the time value of money. The second level introduces the influence of interest which imposes the dimension of time as well as amount in assessing the value of money.

The methods of investment analyses explained in this chapter are used over and over in engineering practice, and in most cases engineers are not required to go beyond these principles. These methods also are the base for extensions into more complex economic analyses.

SEVERAL TEXTS ON ENGINEERING ECONOMICS

Barish, N. N.: "Economic Analysis for Engineering and Managerial Decision-Making," McGraw-Hill Book Company, New York, 1962.

DeGarmo, E. P.: "Engineering Economy," The Macmillan Company, New York, 1967.

Grant, E. L., and W. G. Ireson: "Principles of Engineering Economy," 4th ed., The Ronald Press Company, New York, 1960.

Smith, G. W.: "Engineering Economy," Iowa State University Press, Ames, Iowa, 1968.

Taylor, G. A.: "Managerial and Engineering Economy," D. Van Nostrand Company, Inc., Princeton, N.J., 1964.

PROBLEMS

3-1. Using the digital computer, calculate your personal set of tables for the following factors: CAF, PWF, SCAF, SFF, SPWF, and CRF. Devote a separate page to each of the factors, label adequately, and calculate at the following interest rates: 1, 2, 3, 4, 5, 6, 8, 10, 12, 15, 20, and 25 percent. Calculate for the following interest periods: 1 to 20 by 1s, 22 to 30 by 2s, and 30 to 60 by 5s. Print out the factors to four places after the decimal point.

3-2. Annual investments are being made so that $20,000 will be accumulated at the end of 10 years. The interest rate on these investments is initially expected to be 4 percent, compounded annually. After 4 years, the rate of interest is unexpectedly increased to 5 percent, so the payments for the remaining 6 years can be reduced. What amounts should be invested annually for the first 4 years, and what sums for the last 6?
Ans.: $1,666, $1,547.

3-3. A $1,000 bond was issued 5 years ago and will mature 5 years from now. The bond yields an interest rate of 5 percent or $50 per year. The owner of the bond wishes to sell the bond, but since interest rates have increased a prospective buyer wishes to earn a rate of 6 percent on his investment. What should be the selling price, keeping in mind that the purchaser receives $50 per year, which is reinvested, and receives the $1,000 face value at maturity? Interest is compounded annually.
Ans.: $957.80.

3-4. A firm wishes to set aside equal amounts at the end of each of 10 years, beginning at the end of the first year, in order to have $8,000 maintenance funds available at the end of the seventh, eighth, ninth, and tenth years. What is the required annual payment if the money is invested and draws 6 percent, compounded annually?
Ans.: $2,655.

3-5. A municipality must build a new electric generating plant and can choose between a steam or a hydro facility. The anticipated cost of the steam plant is $10,000,000. Comparative data for the two plants are:

	Generating cost including maintenance per kWh	Expected life, years	Salvage value, %
Steam	$0.004	20	10
Hydro	0.002	30	10

The expected annual consumption of power is 300 million kWh. If money is borrowed at 5 percent interest, compounded annually, what first cost of the hydro plant would make the two alternatives equally attractive investments?
Ans.: $21,570,000.

3-6. Calculate the uniform annual profit on a processing plant for which the following data apply:

Life	12 years
First cost	$280,000
Annual real estate tax and insurance	4 percent of first cost
Salvage value at the end of 12 years	$50,000
Annual cost of raw materials, labor, and other supplies	$60,000
Income	$140,000
Maintenance costs: During first year	0
At end of second year	$1,000
At end of third year	$2,000
. . . .	
At end of twelfth year	$11,000

The interest rate applicable is 6 percent compounded annually.
Ans.: Profit = $33,560.

3-7. A proposed investment consists of constructing a building, purchasing production machinery, and operating for a period of 20 years. The expected life of the building is 20 years, its first cost is $250,000 with a salvage value of $50,000. The maximum life of the machinery is 12 years, so it will be necessary to renew the machinery once during the 20 years. The first cost of the machinery is $132,000 and its salvage value = $132,000/(age, years). The annual income less operating expense is expected to be $50,000. Interest is 6 percent per year, compounded annually.

(a) When is the most favorable time to replace the machinery?

(b) Compute the present worth of the profit.

Ans.: $152,000.

3-8. A processing plant has a first cost of $600,000 and has an expected life of 15 years with no salvage value. Money is borrowed on an 8 percent interest rate and the first cost is paid off with 15 equal annual payments. The expected annual income is $200,000 and annual operating expenses are $40,000. Corporation income tax is 50 percent of the profits before taxes, and the sum-of-the-year's digits method of depreciation is applicable on the tax life of the facility which is 12 years with no salvage value. Compute the income tax for (a) the first year, and (b) the second year.

Ans.: $9,850, $14,580.

4
Equation Fitting and Mathematical Modeling

4-1 Introduction Both equation fitting and mathematical modeling are processes of finding equations that represent properties of substances or performance characteristics of equipment. The distinction between the two is generally that equation fitting starts with some data and uses only the numerical relationship among the data points to arrive at an equation, while mathematical modeling utilizes physical principles applicable to the substance, process, or component to arrive at an equation.

Equation fitting and mathematical modeling are not ends in themselves but are a step toward system simulation and optimization. In order to optimize a system, it is essential to predict the performance of that system in order to evaluate adjustments in component size and performance. System simulation becomes, then, a step in optimization as well as a useful activity in its own right.

To simulate a system, the performance of each component of that system must be known, and at the operating condition of the system all the individual performance characteristics of the components must be satisfied. Steady-state simulation becomes, therefore, a simultaneous solution of the performance characteristics of all components in the system.

Performance characteristics of components can be expressed in one of three forms—graphical, tabular, or equation. The prediction of system performance by finding the intersection of performance curves of the components of a system is a standard engineering procedure, but it is feasible only for small systems with two or three components. Searching through a multiplicity of tables to find the operating conditions that satisfy all of the components is also not practical. The one method left is to express the performance of the components in equation form and solve these equations simultaneously. With the availability of a digital computer, it is possible to solve dozens or even hundreds of simultaneous equations, each representing the performance of a component in the system.

This chapter will discuss some of the systematic procedures used in devising equations to fit available data. Equation fitting, however, is also an art, because intuition and ingenuity are valuable in sensing the form of the equation to be chosen. Some of the most widely used techniques that will be discussed in this chapter are polynomial and exponential representations. Because of the frequency with which heat exchangers and turbomachinery appear in thermal systems, a brief review will be made of the performance equations of these types of equipment.

4-2 Component simulation A typical situation where the performance of a component is to be modeled is where an equation is sought that expresses the mass rate of flow delivered by an axial-flow air compressor as a function of the inlet pressure and temperature, the speed, and the outlet pressure. Another example is the representation of the cost of a pump as a function of its size. It is assumed that tabular or graphical data are available from manufacturer's catalogs for the component. It is further understood that we shall take advantage of all physical laws or mathematical relations that are available. For example, in simulating the performance of a counter-flow heat exchanger, we shall take advantage of the knowledge that the potential for heat transfer is the log-mean temperature difference. It is unnecessary to process reams of test data with dozens of combinations of inlet and outlet temperatures to seek correlations of the temperature patterns.

Equation fitting is to a large extent an art. There are countless forms of equations, for example, polynomials, exponentials, and trigonometric functions, and there is no methodical procedure for selecting the most applicable one. Plotting a graph of the function is often a useful first step to explore its nature. Usually there is more than one possible representation and the criterion of success is how well the equation fits the available data.

4-3 Polynomial representations Probably the most obvious and most useful form of representation is a polynomial. If y is to be represented as a function of x, the polynomial form is

$$y = a_0 + a_1x + a_2x^2 + \cdots + a_nx^n \tag{4-1}$$

where a_0 to a_n are constants. The degree of the equation is the highest exponent of x, which in Eq. (4-1) is n.

Equation (4-1) is an expression giving the function of one variable in terms of another. Other common situations are where one variable is a function of two or more variables, for example, in the axial-flow compressor mentioned in Sec. 4-2.

$$\text{Flow rate} = f(\text{inlet pressure, inlet temperature, compressor}$$
$$\text{speed, outlet pressure})$$

When the number of data points available is precisely the degree of the equation plus 1, $n + 1$, a polynomial can be devised that exactly expresses those data points. When the number of available data points exceeds $n + 1$, it may be advisable to seek a polynomial that gives the "best fit" to the data points.

The first and simplest case to be considered will be where one variable is a function of another variable and where the number of data points equals $n + 1$.

4-4 Polynomial, one variable a function of another variable and $n + 1$ data points Two available data points are adequate to describe a first-degree or linear equation such as shown in Fig. 4-1. The form of this first-degree equation is

$$y = a_0 + a_1x \tag{4-2}$$

The xy pairs for the two known points (x_0, y_0) and (x_1, y_1) can be substituted into Eq. (4-2), providing two linear equations with two unknowns—a_0 and a_1.

$$y_0 = a_0 + a_1x_0$$
$$y_1 = a_0 + a_1x_1$$

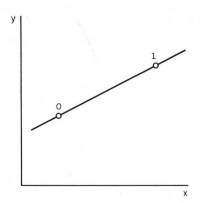

Fig. 4-1 Two points describing a linear equation.

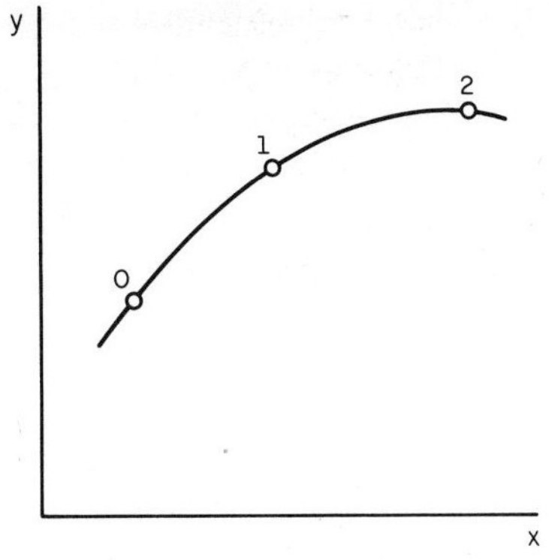

Fig. 4-2 Three points describing a quadratic equation.

For a second-degree or quadratic equation, three data points are needed, for example, points 0, 1, and 2 in Fig. 4-2. The xy pairs for the three known points can be substituted into the general form for the quadratic equation

$$y = a_0 + a_1 x + a_2 x^2 \tag{4-3}$$

giving

$$y_0 = a_0 + a_1 x_0 + a_2 x_0{}^2$$
$$y_1 = a_0 + a_1 x_1 + a_2 x_1{}^2$$
$$y_2 = a_0 + a_1 x_2 + a_2 x_2{}^2$$

The solution of these three linear, simultaneous equations provides the values of a_0, a_1, and a_2.

The coefficients of the high-degree terms in a polynomial may be quite small, particularly if the independent variable is large. For example, if the enthalpy of saturated water vapor h is a function of temperature t in the equation

$$h = a_0 + a_1 t + \cdots + a_5 t^5 + a_6 t^6$$

where the range of t extends into hundreds of degrees, the value of a_5 and a_6 may be so small that precision problems result. Sometimes this difficulty can be surmounted by defining a new independent variable, for example, $t/100$.

$$h = a_0 + a_1 \left(\frac{t}{100}\right) + \cdots + a_5 \left(\frac{t}{100}\right)^5 + a_6 \left(\frac{t}{100}\right)^6$$

4-5 Solving simultaneous, linear equations—Gaussian elimination

The need for solving linear, simultaneous equations was encountered in the preceding section, and it will not be the last time this need is experienced in the study of thermal systems. It may be profitable to pause to discuss one method for solving simultaneous equations. Gaussian elimination is a method that lends itself conveniently to a computer solution, but the reader who has a program or library routine using other methods may bypass this section. However, it is useful to have a program for personal use that can be applied as a subroutine in the future.

Equations (4-4) to (4-6) are a set of linear equations that will be solved by Gaussian elimination.

$$x_1 - 4x_2 + 3x_3 = -7 \tag{4-4}$$

$$3x_1 + x_2 - 2x_3 = 14 \tag{4-5}$$

$$2x_1 + x_2 + x_3 = 5 \tag{4-6}$$

The two major steps in Gaussian elimination are:

1. Conversion of the coefficient matrix into a triangular matrix
2. Solution for x_n to x_1 by back substitution

In the example set of equations, the first part of step 1 is to eliminate the coefficients of x_1 in Eq. (4-5) by multiplying Eq. (4-4) by a suitable constant and adding the product to Eq. (4-5). Specifically, multiply Eq. (4-4) by -3 and add to Eq. (4-5). Similarly, multiply Eq. (4-4) by -2 and add to Eq. (4-6).

$$x_1 - 4x_2 + 3x_3 = -7 \tag{4-7}$$

$$13x_2 - 11x_3 = 35 \tag{4-8}$$

$$9x_2 - 5x_3 = 19 \tag{4-9}$$

The last part of step 1 is to multiply Eq. (4-8) by $-\frac{9}{13}$ and add to Eq. (4-9), which completes the triangularization.

$$x_1 - 4x_2 + 3x_3 = -7 \tag{4-10}$$

$$13x_2 - 11x_3 = 35 \tag{4-11}$$

$$\frac{34}{13}x_3 = -\frac{68}{13} \tag{4-12}$$

Moving now to step 2, the value of x_3 can be determined directly from Eq. (4-12), $x_3 = -2$. Substituting this value of x_3 into Eq.

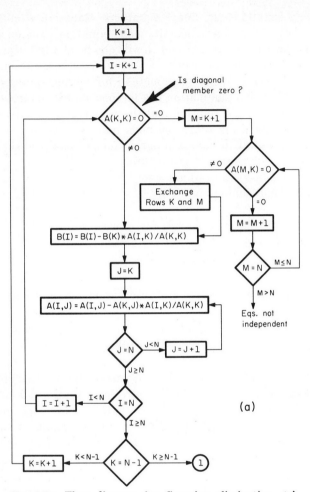

Fig. 4-3a Flow diagram for Gaussian elimination—triangularization portion.

(4-11) and solving, $x_2 = 1$.   Finally, substitute the values of x_2 and x_3 into Eq. (4-10) to find that $x_1 = 3$.

Next using the form adaptable to a digital computer, the set of N simultaneous equations can be written in a FORTRAN-like form

$$A(1,1)*X(1) + A(1,2)*X(2) + \cdots + A(1,N)*X(N) = B(1)$$
$$A(2,1)*X(1) + A(2,2)*X(2) + \cdots + A(2,N)*X(N) = B(2)$$
$$\cdots \cdots \cdots \cdots \cdots \cdots \cdots \cdots \cdots \cdots \cdots \cdots \cdots$$
$$A(N,1)*X(1) + A(N,2)*X(2) + \cdots + A(N,N)*X(N) = B(N)$$

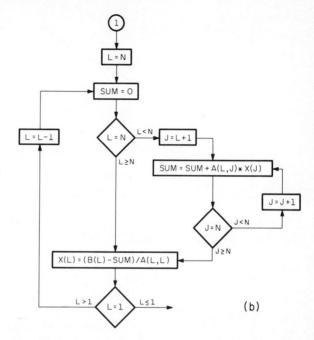

Fig. 4-3*b*   Flow diagram for Gaussian elimination—back substitution portion.

During the elimination process of step 1, the equation

$$A(I,J) = A(I,J) - A(K,J)*A(I,K)/A(K,K) \qquad (4\text{-}13)$$

loops through the I and J indices.   The index K starts at 1 and proceeds to N as the elimination proceeds down the diagonal.   One precaution that must be observed is that $A(K,K)$ could conceivably be zero which would invalidate Eq. (4-13).   In Eq. (4-8), for example, it would have been possible with a different set of equations for the coefficient of x_2 to be zero instead of 13.   If that had occurred, Eqs. (4-8) and (4-9) could have been interchanged, and then Eq. (4-13) would again have been applicable.   If both of the x_2 coefficients in Eqs. (4-8) and (4-9) had been zero, this would indicate that the equations in the set are not independent.

A possible flow diagram for the calculation portion without the input-output instructions for the Gaussian elimination is shown in Fig. 4-3.   The diagram is applicable to a set of N equations.

4-6   Simplifications when independent variable is uniformly spaced
Sometimes a polynomial is used to represent a function, $y = f(x)$,

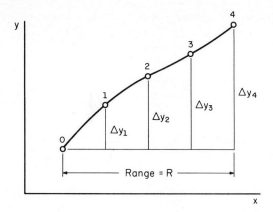

Fig. 4-4 Polynomial representation when points are equally spaced along the x axis.

for example, where the values of y are known at equally spaced values of x. This situation exists, for instance, when the data points are read off a graph where the points can be chosen at equal intervals of x. The solution of simultaneous equations to determine the coefficients in the polynomial can be performed symbolically in advance,[1] and the execution of the calculations requires a relatively small effort thereafter.

Suppose that the curve in Fig. 4-4 is to be reproduced by a fourth-degree polynomial. The $n + 1$ data points (five in this case) establish a polynomial of degree n (four in this case). The spacing of the points is $x_1 - x_0 = x_2 - x_1 = x_3 - x_2 = x_4 - x_3$. The range of x, $x_4 - x_0$ is designated R, and the symbols are $\Delta y_1 = y_1 - y_0$; $\Delta y_2 = y_2 - y_0$, etc.

Instead of the polynomial form of Eq. (4-1), an alternate form is used.

$$y - y_0 = a_1 \left[\frac{n}{R} (x - x_0) \right] + a_2 \left[\frac{n}{R} (x - x_0) \right]^2$$
$$+ a_3 \left[\frac{n}{R} (x - x_0) \right]^3 + a_4 \left[\frac{n}{R} (x - x_0) \right]^4 \quad (4\text{-}14)$$

To find a_1 to a_4, first substitute the (x_1, y_1) pair into Eq. (4-14).

$$\Delta y_1 = a_1 \left[\frac{4(x_1 - x_0)}{R} \right] + a_2 \left[\frac{4(x_1 - x_0)}{R} \right]^2$$
$$+ a_3 \left[\frac{4(x_1 - x_0)}{R} \right]^3 + a_4 \left[\frac{4(x_1 - x_0)}{R} \right]^4 \quad (4\text{-}15)$$

[1] See M. W. Wambsganss, Jr., Curve Fitting with Polynomials, *Mach. Des.*, vol. 35, no. 10, p. 167, April 25, 1963.

Because of the uniform spacing of the points along the x axis, $n(x_1 - x_0)/R = 1$, and so Eq. (4-15) can be rewritten as

$$\Delta y_1 = a_1 + a_2 + a_3 + a_4 \tag{4-16}$$

Using the (x_2, y_2) pair and the fact that $n(x_2 - x_0)/R = 2$, gives

$$\Delta y_2 = 2a_1 + 4a_2 + 8a_3 + 16a_4 \tag{4-17}$$

Similarly, for (x_3, y_3) and (x_4, y_4),

$$\Delta y_3 = 3a_1 + 9a_2 + 27a_3 + 81a_4 \tag{4-18}$$

$$\Delta y_4 = 4a_1 + 16a_2 + 64a_3 + 256a_4 \tag{4-19}$$

The expressions for a_1 to a_4, found by solving Eqs. (4-16) to (4-19) simultaneously, are shown in Table 4-1, along with the constants for the cubic (see Prob. 4-1), quadratic, and linear equations.

4-7 Lagrange interpolation Another form of polynomial is that which results when using Lagrange interpolation. This method is applicable, unlike the method described in Sec. 4-6, to arbitrary spacing along the x axis, it has the advantage of not requiring the simultaneous solution of equations, but suffers the disadvantage of being cumbersome to write out. The disadvantage is not applicable if the calculation is performed on a digital computer, in which case the programming is quite compact.

Using a quadratic equation as an example, the usual form for a function of one variable is

$$y = a_0 + a_1 x + a_2 x^2 \tag{4-20}$$

Table 4-1   Constants in Eq. (4-14)

Equation	a_4	a_3	a_2	a_1
Fourth degree	$\frac{1}{24}(\Delta y_4 - 4\Delta y_3 + 6\Delta y_2 - 4\Delta y_1)$	$\dfrac{\Delta y_3}{6} - \dfrac{\Delta y_2}{2} + \dfrac{\Delta y_1}{2} - 6a_4$	$\dfrac{\Delta y_2}{2} - \Delta y_1 - 3a_3 - 7a_4$	$\Delta y_1 - a_2 - a_3 - a_4$
Cubic		$\frac{1}{6}(3\Delta y_1 + \Delta y_3 - 3\Delta y_2)$	$\frac{1}{2}(\Delta y_2 - 2\Delta y_1) - 3a_3$	$\Delta y_1 - a_2 - a_3$
Quadratic			$\frac{1}{2}(\Delta y_2 - 2\Delta y_1)$	$\Delta y_1 - a_2$
Linear				Δy_1

For Lagrange interpolation, a revised form is used

$$y = c_1(x - x_2)(x - x_3) + c_2(x - x_1)(x - x_3)$$
$$+ c_3(x - x_1)(x - x_2) \quad (4\text{-}21)$$

The three available data points are (x_1, y_1), (x_2, y_2), and (x_3, y_3). Equation (4-21) could be multiplied out and terms collected to show the correspondence to the form in Eq. (4-20).

By setting $x = x_1$, x_2, and x_3 in turn in Eq. (4-21), the constants can be found quite simply.

$$c_1 = \frac{y_1}{(x_1 - x_2)(x_1 - x_3)}$$

$$c_2 = \frac{y_2}{(x_2 - x_1)(x_2 - x_3)}$$

$$c_3 = \frac{y_3}{(x_3 - x_1)(x_3 - x_2)}$$

The general form of the equation for finding the value of y for a given x when n data points are known is

$$y = \sum_{i=1}^{n} y_i \prod_{j=1}^{n} \frac{(x - x_j)}{(x_i - x_j)} \quad \frac{\text{omitting } (x - x_i)}{\text{omitting } (x_i - x_i)} \quad (4\text{-}22)$$

where the symbol pi indicates multiplication.

The equation represented by Eq. (4-22) is a polynomial of degree $n - 1$.

The flow diagram, with the $I = 1$ to N pairs of $X(I)$ and $Y(I)$ points already stored, for the calculation of y (designated WHY) at a desired value of x (designated EX) is shown in Fig. 4-5.

Equation (4-21) is the quadratic form for a function of one variable. The Lagrange interpolation is applicable to functions of more than one variable. As an example, the quadratic form of a function of two variables

$$z = a_1 + a_2 x + a_3 y + a_4 xy + a_5 x^2 + a_6 y^2 + a_7 xy^2$$
$$+ a_8 x^2 y + a_9 x^2 y^2 \quad (4\text{-}23)$$

can be rewritten in the form convenient for Lagrange interpolation

$$z = c_{11}(x - x_2)(x - x_3)(y - y_2)(y - y_3)$$
$$+ c_{12}(x - x_2)(x - x_3)(y - y_1)(y - y_3)$$
$$+ c_{13}(x - x_2)(x - x_3)(y - y_1)(y - y_2)$$
$$+ \cdots + c_{33}(x - x_1)(x - x_2)(y - y_1)(y - y_2) \quad (4\text{-}24)$$

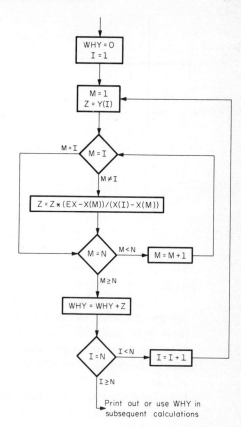

Fig. 4-5 Flow diagram for Lagrange interpolation.

Print out or use WHY in subsequent calculations

Nine values of z must be known at the combination of x's and y's, as shown in Fig. 4-6. As an example of determining the constants in Eq. (4-24), the constant c_{12} can be determined by the equation

$$c_{12} = \frac{z_{12}}{(x_1 - x_2)(x_1 - x_3)(y_2 - y_1)(y_2 - y_3)} \tag{4-25}$$

4-8 Best fit—the method of least squares When a polynomial of degree n is to be found where more than $n + 1$ data points are available, it would be an exceptional case if the polynomial accurately represented all of the data points. It is possible that the polynomial that has the least total deviation from the data points may not pass through any of the points. One definition of a *best-fit* curve is the one where the sum of the absolute values of the deviations from the data points is a minimum. Another type of best-fit curve, which is slightly different from the one just mentioned, is the

	y_1	y_2	y_3
x_1	z_{11}	z_{12}	z_{13}
x_2	z_{21}	z_{22}	z_{23}
x_3	z_{31}	z_{32}	z_{33}

Fig. 4-6 Data points for Lagrange interpolation.

one where the sum of the squares of the deviation is a minimum. The procedure in determining this curve or polynomial is called the *method of least squares.*

Some people proudly announce their use of the method of least squares, in order to emphasize the care that they have lavished on their data analysis. Misuses of the method, as illustrated in Fig. 4-7a and b are not uncommon. In Fig. 4-7a, while a straight line can be found that results in the least-squares deviation, the correlation between the x and y variables seems questionable, and perhaps no such device can improve the correlation. The scatter may be due to the omission of some significant variable(s). In Fig. 4-7b, it would have been preferable to eyeball in the curve, rather than to fit a straight line to the data by the least-squares method. The error was not in using least squares, but in applying a curve of too

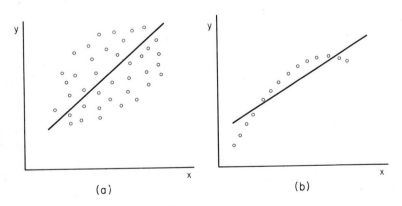

Fig. 4-7 Misuses of the method of least squares.

low a degree. The least-squares method also applies to polynomials of a degree higher than 1.

The procedure for using the least-squares method for first- and second-degree polynomials will be explained here. Consider first the linear equation of the form

$$y = a + bx \tag{4-26}$$

where m pairs of data points are available: (x_1, y_1), (x_2, y_2), . . . , (x_m, y_m). The deviation of the data point from that calculated from the equation is $(a + bx_i - y_i)$. We wish to choose an a and b such that the summation

$$\sum_{i=1}^{m} (a + bx_i - y_i)^2 \rightarrow \text{minimum} \tag{4-27}$$

The minimum occurs when the partial derivatives of Eq. (4-27) with respect to a and b equal zero.

$$\frac{\partial \sum_{i=1}^{m} (a + bx_i - y_i)^2}{\partial a} = \sum 2(a + bx_i - y_i) = 0$$

and

$$\frac{\partial \sum_{i=1}^{m} (a + bx_i - y_i)^2}{\partial b} = \sum 2(a + bx_i - y_i)x_i = 0$$

Dividing by 2 and separating the above two equations into individual terms,

$$ma + b \sum x_i = \sum y_i \tag{4-28}$$

$$a \sum x_i + b \sum x_i^2 = \sum x_i y_i \tag{4-29}$$

Solving Eqs. (4-28) and (4-29) simultaneously yields the expressions for a and b

$$a = \frac{\sum x_i^2 \sum y_i - \sum x_i \sum x_i y_i}{m \sum x_i^2 - \left(\sum x_i\right)^2} \tag{4-30}$$

and

$$b = \frac{m \sum x_i y_i - \sum x_i^2 \sum y_i}{m \sum x_i^2 - \left(\sum x_i\right)^2} \tag{4-31}$$

A similar procedure can be followed when fitting a parabola of the form

$$y = a + bx + cx^2 \tag{4-32}$$

to m data points.   The summation to be minimized is

$$\sum_{i=1}^{m} (a + bx_i + cx_i^2 - y_i)^2 \rightarrow \text{minimum}$$

Differentiating partially with respect to a, b, and c, in turn, results in three linear, simultaneous equations

$$ma + b \sum x_i + c \sum x_i^2 = \sum y_i \tag{4-33}$$

$$a \sum x_i + b \sum x_i^2 + c \sum x_i^3 = \sum x_i y_i \tag{4-34}$$

$$a \sum x_i^2 + b \sum x_i^3 + c \sum x_i^4 = \sum x_i^2 y_i \tag{4-35}$$

If the equation sought is one of first degree, then possibly the coefficients a and b would be calculated by hand using Eqs. (4-30) and (4-31).   For a second-degree representation requiring the solution of three simultaneous equations, the solution might be performed on a computer, and surely this would be true when fitting equations of higher degree.   The summations, then, become the coefficients which are entered into the computer program.   Furthermore, a pattern can be observed by comparing the two sets of equations, Eqs. (4-28) to (4-29) and Eqs. (4-33) to (4-35).   The coefficient matrix contains identical terms in the diagonals that run upward to the right.   These coefficients are summations of progressively higher powers to x_i as one moves down the matrix. The constants on the right side of the equalities are summations of products of x_i and y_i with the power of x_i increasing down the column.

4-9   Exponential forms   The dependence of one variable on a second variable raised to a power is a frequent physical relation and thus occurs often in engineering practice.   The graphical method of determining the constants b and m in the equation $y = bx^m$ by graphical means is the simplest example of mathematical modeling using exponential forms.   On a graph of the known values of x and y on a log-log plot, Fig. 4-8, the slope of the straight line through the points equals m, and the intercept at $x = 1$ defines $\log b$.

Sections 4-10 and 4-11 explore extensions to the simple exponential form.

4-10   Exponential with a constant   An extension of the equation discussed in Sec. 4-9 introduces a constant b in the linear portion of the equation to generate the form

$$y - b = ax^m \tag{4-36}$$

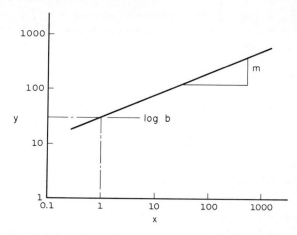

Fig. 4-8 Graphical determination of the constant b and exponent m.

This equation permits representations of curves similar to those shown in Fig. 4-9. The curve shown in Fig. 4-9b is especially frequent in engineering practice. The function y approaches some value b asymptotically as x increases.

The steps in a graphical method for determining a, b, and m in Eq. (4-36) when pairs of xy values are known is as follows:

1. Estimate the value of b.
2. Use the steep portion of the curve to evaluate m by a log-log plot of $(y - b)$ versus x in a manner similar to that in Sec. 4-9.

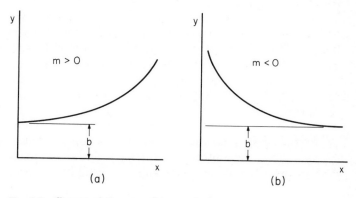

Fig. 4-9 Curves of the equation $y - b = ax^m$.

Table 4-2   Motor costs in Example 4-1

hp	Cost, $	$/hp
0.50	50	100.00
0.75	60	80.00
1.00	70	70.00
1.50	90	60.00
2.00	110	55.00
3.00	150	50.00
5.00	220	44.00
7.50	305	40.50
15.00	560	37.30

3. With the value of m from step 2, plot a graph of y versus x^m. The resulting curve should be a straight line with a slope of "a" and an intercept that indicates a more correct value of b.

Example 4-1 Using the data provided in Table 4-2, express the cost-per-unit capacity in $/hp for electric motors, in an equation of the form

$$\$/\text{hp} = b + a(\text{hp})^m$$

Solution The costs shown in the table are typical of most engineering equipment in that the cost per unit capacity decreases as the capacity increases until some minimum is reached.

Using the three steps to find the constants, first estimate from the data that the $/hp levels out at a value of about 32, which becomes the trial value of b.

In step 2, plot the values of ($/hp $- 32$) versus hp for the steep part of the curve, as in Fig. 4-10, since the steep part of the curve in Fig. 4-9b controls the value of m. The small motor sizes are the ones in the steep portion of the curve, and from the slope of the line in Fig. 4-10 the value of m is found to be -0.865.

In step 3, the $/hp values are plotted against $(\text{hp})^{-0.865}$, as in Fig. 4-11. A straight line through the points has a slope of 36 and an intercept of 34.5. The magnitude of "a" is therefore 36, and the more correct value of b is 34.5. The resulting equation, then, is

$$\$/\text{hp} = 34.50 + \frac{36}{(\text{hp})^{0.865}}$$

It may be possible to fit a curve by combining two or more forms. For example, in Fig. 4-12, suppose that the value of y approaches asymptotically a straight line as x increases. A reasonable approach to this modeling task would be to propose that

$$y = y_1 + y_2 = (a + bx) + (c + dx^m)$$

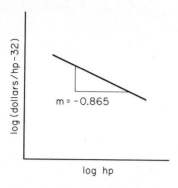

Fig. 4-10 Determination of exponent m.

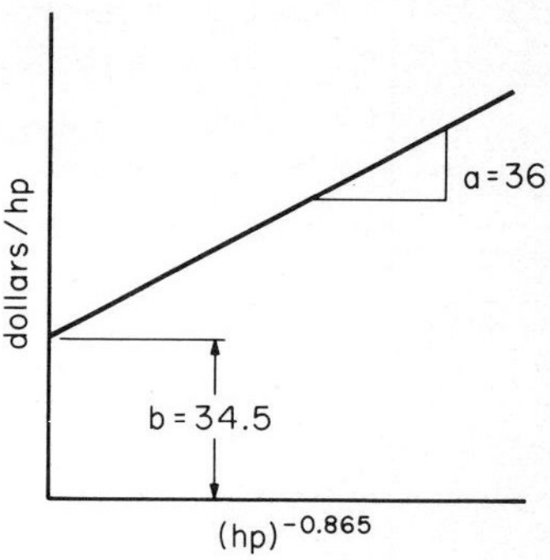

Fig. 4-11 Determination of a and b.

where y_1 is the straight line that is the asymptote of the curve, and y_2 is of the form of Eq. (4-36).

4-11 Function of two variables An operating variable of a component is often a function of two other variables—not just one. For example, the head developed by the centrifugal pump shown in Fig. 4-13 is a function of both the speed n and flow rate w.

If a polynomial expression for the head h is sought in terms of a second-degree equation in n and w, separate equations may be written for each of the three curves in Fig. 4-13. Three points on the 1,750-rpm curve would provide the constants in the equation

$$h_1 = a_1 + b_1 w + c_1 w^2 \tag{4-37}$$

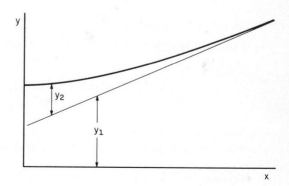

Fig. 4-12 Combination of two forms.

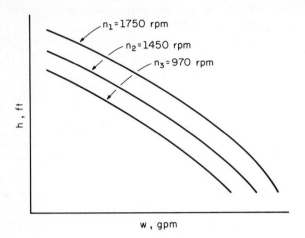

Fig. 4-13   Performance of a centrifugal pump.

Similar equations for the curves for the 1,450 rpm and 970 rpm speeds are

$$h_2 = a_2 + b_2 w + c_2 w^2 \tag{4-38}$$

$$h_3 = a_3 + b_3 w + c_3 w^2 \tag{4-39}$$

Next the a constants may be expressed as a second-degree equation in terms of n, using the three data points $(a_1, 1{,}750)$, $(a_2, 1{,}450)$, and $(a_3, 970)$.   Such an equation would have the form

$$a = A_0 + A_1 n + A_2 n^2 \tag{4-40}$$

Similarly, for b and c,

$$b = B_0 + B_1 n + B_2 n^2 \tag{4-41}$$

$$c = C_0 + C_1 n + C_2 n^2 \tag{4-42}$$

Finally, place the constants of Eqs. (4-40) to (4-42) into the general equation

$$h = A_0 + A_1 n + A_2 n^2 + (B_0 + B_1 n + B_2 n^2)w \\ + (C_0 + C_1 n + C_2 n^2)w^2 \tag{4-43}$$

The A, B, and C constants can all be assigned numerical values if nine data points from Fig. 4-13 are available.

Example 4-2   Manufacturers of cooling towers often present catalog data showing the outlet-water temperature as a function of the wet-bulb temperature of the ambient air and the *range*.   The range is the difference

Table 4-3 Outlet-water temperature of cooling tower in Example 4-2

Range, °F	Wet-bulb temperature, °F		
	70	74	78
20	80.0	82.8	85.4
30	82.0	84.5	87.2
40	83.8	85.9	88.6

between the inlet and the outlet temperature. In Table 4-3, for example, when the wet-bulb temperature is 70°F and the range is 20°F, the temperature of the leaving water is 80°F, so the temperature of the entering water is $80 + 20 = 100°F$. Express the outlet-water temperature t in Table 4-3 as a function of the wet-bulb temperature WBT and the range R.

Solution Choose second-degree representations for both the independent variables. With a 70°F wet-bulb temperature, the three pairs of points (20, 80.0), (30, 82.0), and (40, 83.8) can be represented by a parabola

$$t = 75.4 + 0.25R - 0.001R^2$$

When the WBT = 74°F,

$$t = 78.5 + 0.245R - 0.0015R^2$$

When the WBT = 78°F,

$$t = 80.6 + 0.28R - 0.002R^2$$

Next, the numerical terms 75.4, 78.5, and 80.6 can be expressed as a parabola in terms of the WBT

$$-140.7 + 5.275 \text{ WBT} - 0.03125 \text{ WBT}^2$$

The coefficients of R and R^2 can also be expressed in terms of the WBT providing the complete equation

$$\begin{aligned}
t = &(-140.7 + 5.275 \text{ WBT} - 0.03125 \text{ WBT}^2) \\
&+ (6.8125 - 0.18125 \text{ WBT} + 0.00125 \text{ WBT}^2)R \\
&+ (0.00775 - 0.000125 \text{ WBT})R^2
\end{aligned}$$

The previous discussion was based on polynomial representations. Other forms of representations, such as exponential, also submit to the same procedure of determining a set of constants for equations at given values of one variable, then finding equations for these constants as a function of the second variable.

4-12 Factorial plan When there is a reasonable expectation that a function can be represented as the sum of the functions of the individual independent variables, the factorial plan provides an effective method of representation. The factorial plan applies, for example, to a function $w(x, y, z)$ if the form of the function is

$$w = f_1(x) + f_2(y) + f_3(z) \tag{4-44}$$

The functions f_1, f_2, and f_3 may be complicated ones, but it is necessary that f_1 be a function of x only, f_2 of y only, and f_3 of z only.

Of course, the functions could be combined with negative signs in Eq. (4-44), and it is even possible for the functions to be products, such as,

$$u = f_4(x)f_5(y)f_6(z) \tag{4-45}$$

because Eq. (4-45) could be rewritten as the sum of logarithms

$$\log u = \log [f_4(x)] + \log [f_5(y)] + \log [f_6(z)]$$

Returning to the representation of Eq. (4-44), in order for the factorial plan to apply the values of w must be known at combinations of values x, y, and z as indicated by Fig. 4-14. The symbols x_1, x_2, and x_3, for example, represent three different values of x. The value of w in the upper-left block is called w_a and in terms of the independent variables is

$$w_a = f_1(x_3) + f_2(y_1) + f_3(z_1) \tag{4-46}$$

Similarly,

$$w_b = f_1(x_3) + f_2(y_2) + f_3(z_2) \tag{4-47}$$

$$w_c = f_1(x_3) + f_2(y_3) + f_3(z_3) \tag{4-48}$$

	y_1	y_2	y_3
x_3	(a) z_1	(b) z_2	(c) z_3
x_2	(d) z_2	(e) z_3	(f) z_1
x_1	(g) z_3	(h) z_1	(i) z_2

Fig. 4-14 Factorial plan.

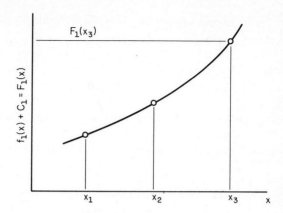

Fig. 4-15 The function $f_1(x) + C_1$.

Summing Eqs. (4-46) to (4-48) and solving for $f_1(x_3)$,

$$f_1(x_3) = \frac{w_a + w_b + w_c - \sum_{i=1}^{3} f_2(y_i) - \sum_{i=1}^{3} f_3(z_i)}{3} \tag{4-49}$$

Designating the two summations in Eq. (4-49) as $3C_1$,

$$f_1(x_3) + C_1 = \frac{w_a + w_b + w_c}{3} = F_1(x_3) \tag{4-50}$$

where $F_1(x)$ is defined as $f_1(x) + C_1$. Equation (4-50) provides the $F_1(x_3)$ point on the graph of Fig. 4-15.

The summation of the w values for the second row in Fig. 4-14 provides $F_1(x_2)$, and the bottom row provides $F_1(x_1)$. Thus, as Fig. 4-15 shows, the function F_1 has been determined and this function differs from f_1 by only the constant C_1.

The function F_2, which is defined as f_2 plus a constant C_2, can be determined by summations of the w values in the columns of Fig. 4-14. Next, the function f_3 plus a constant C_3 can be found by summations of the w values: first, of w_a, w_f, and w_h to determine $F_3(z_1)$; then, w_b, w_d, and w_i; and finally, w_c, w_e, and w_g.

The three functions, F_1, F_2, and F_3 are now known, but their sum differs from the sum of f_1, f_2, and f_3 by the sum of the three constants C_1, C_2, and C_3, which can be lumped into one constant C_4. The magnitude of C_4 can be found by substituting any one of the original data points into the equation

$$w(x_i, y_j, z_k) = F_1(x_i) + F_2(y_j) + F_3(z_k) - C_4 \tag{4-51}$$

Equation (4-51), with the known value of C_4, then becomes the

usable expression for w in a slightly revised form compared to Eq.
(4-48).

The factorial plan only separates the effects of the functions
F_1, F_2, and F_3.   Determining the equation for F_1 as a function of x
is still a subsequent step.

The factorial plan is frequently used in the design of experi-
ments.   If the variables x, y, and z are expected to influence w,
tests are run at values according to Fig. 4-14, which provide the
maximum amount of knowledge for the minimum number of test
runs.

4-13 Heat exchangers Most thermal systems include heat
exchangers, so expressing the performance of an existing heat
exchanger is a frequent requirement.   A typical situation encountered in system simulation is to calculate the outlet-fluid temperatures and the rate of heat transfer when the area, overall heat-
transfer coefficient, and inlet temperatures of the two fluids are
known.

The relations for a counterflow heat exchanger with symbols
as shown in Fig. 4-16, will be presented first.

Three equations for the rate of heat transfer q Btu/hr are

$$q = W_1(t_1 - t_2) \tag{4-52}$$

$$q = W_2(t_0 - t_i) \tag{4-53}$$

and

$$q = UA \frac{(t_1 - t_0) - (t_2 - t_i)}{\ln\left[(t_1 - t_0)/(t_2 - t_i)\right]} \tag{4-54}$$

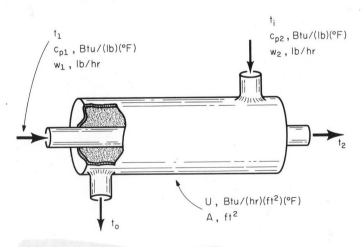

Fig. 4-16   Counterflow heat exchanger.

where

$$W_1 = w_1 c_{p1} \quad \text{and} \quad W_2 = w_2 c_{p2}$$

Equations (4-52) to (4-54) contain three unknowns, q, t_0, and t_2. Reducing the number of equations to two by eliminating q,

$$W_1(t_1 - t_2) = W_2(t_0 - t_i) \tag{4-55}$$

$$W_1(t_1 - t_2) = UA \frac{(t_1 - t_0) - (t_2 - t_i)}{\ln[(t_1 - t_0)/(t_2 - t_i)]} \tag{4-56}$$

Solving for t_0 in Eq. (4-55) and substituting into Eq. (4-56),

$$\ln \frac{t_1 - [t_i + (W_1/W_2)(t_1 - t_2)]}{t_2 - t_i} = UA \left(\frac{1}{W_1} - \frac{1}{W_2} \right)$$

Define D as

$$D = UA \left(\frac{1}{W_1} - \frac{1}{W_2} \right)$$

Then

$$\frac{t_1 - t_i - (W_1/W_2)(t_1 - t_2)}{t_2 - t_i} = e^D$$

Solving for t_2,

$$t_2 = \frac{t_1(W_1/W_2 - 1) + t_i(1 - e^D)}{W_1/W_2 - e^D}$$

and

$$t_2 = t_1 - (t_1 - t_i) \left(\frac{1 - e^D}{W_1/W_2 - e^D} \right) \tag{4-57}$$

The value of t_2 calculated from Eq. (4-57) can be substituted back into Eq. (4-52) to determine q and into Eq. (4-55) to compute t_0.

For the special case where $W_1 = W_2$, $D = 0$ and Eq. (4-57) becomes indeterminate. To surmount this situation, express D as

$$D = \frac{UA}{W_1} \left(1 - \frac{W_1}{W_2} \right)$$

Next express the exponentials in the indeterminate part of Eq. (4-57) in a series

$$\frac{1 - \left\{ 1 + \dfrac{UA}{W_1}\left(1 - \dfrac{W_1}{W_2}\right) + \dfrac{1}{2}\left[\dfrac{UA}{W_1}\left(1 - \dfrac{W_1}{W_2}\right)\right]^2 + \cdots \right\}}{\dfrac{W_1}{W_2} - \left\{ 1 + \dfrac{UA}{W_1}\left(1 - \dfrac{W_1}{W_2}\right) + \dfrac{1}{2}\left[\dfrac{UA}{W_1}\left(1 - \dfrac{W_1}{W_2}\right)\right]^2 + \cdots \right\}}$$

Cancel where possible and divide both the numerator and denominator by $1 - W_1/W_2$.

$$\frac{-\dfrac{UA}{W_1} - \dfrac{1}{2}\left(\dfrac{UA}{W_1}\right)^2\left(1 - \dfrac{W_1}{W_2}\right) + \cdots}{-1 - \dfrac{UA}{W_1} - \dfrac{1}{2}\left(\dfrac{UA}{W_1}\right)^2\left(1 - \dfrac{W_1}{W_2}\right) + \cdots}$$

Finally, let $W_1 \to W_2$, and call this common value W. Then

$$t_2 = t_1 - \frac{(UA/W)(t_1 - t_i)}{1 + UA/W}$$

or

$$t_2 = t_1 - \frac{t_1 - t_i}{W/UA + 1} \tag{4-58}$$

Physically, when $W_1 = W_2$, the same temperature difference prevails at all parts of the heat exchanger, and this temperature difference is $t_1 - t_0 = t_2 - t_i$.

4-14 Evaporators and condensers

In an evaporator or condenser, as shown in Fig. 4-17, with no superheat or subcooling of the fluid that changes phase, the condensing or evaporating fluid remains at a constant temperature, provided that its pressure does not change.

The heat-transfer equation becomes

$$q = UA\,\frac{(t_c - t_1) - (t_c - t_2)}{\ln\left[(t_c - t_1)/(t_c - t_2)\right]} = W(t_2 - t_1) \tag{4-59}$$

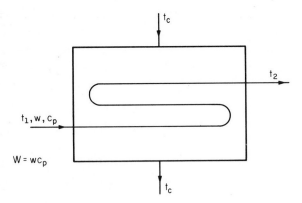

Fig. 4-17 An evaporator or condenser where one fluid remains at a constant temperature.

Equation (4-59) can be converted to the form

$$\frac{UA}{W} = \ln \frac{t_c - t_1}{t_c - t_2} = -\ln \frac{t_c - t_2}{t_c - t_1}$$

Taking the antilog,

$$e^{-UA/W} = \frac{t_c - t_2}{t_c - t_1} = \frac{t_c - t_2 + t_1 - t_1}{t_c - t_1}$$

Then

$$t_2 = t_1 + (t_c - t_1)(1 - e^{-UA/W}) \tag{4-60}$$

4-15 Heat exchange effectiveness The equations that were developed in Secs. 4-13 and 4-14 apply to the counterflow heat exchanger, except for the condenser or evaporator in which case the flow configuration is immaterial.   In many commercial heat exchangers, due to return-tube circuiting, combinations of counterflow, parallel flow, and cross flow exist.   To calculate the outlet conditions of such heat exchangers, *effectiveness* offers great convenience.   The effectiveness in words is defined as the ratio of the actual rate of heat transfer to the maximum possible if the heat exchanger were of the counterflow type with infinite area.   Referring to the counterflow heat exchanger in Fig. 4-16, if the area were infinite the temperature distributions would appear as in either Fig. 4-18a or b, depending upon the relative values of W_{cold} and W_{hot}, where $W = wc_p$.

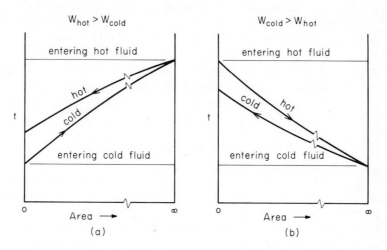

Fig. 4-18 Temperature distributions in a counterflow heat exchanger of infinite area.

The maximum possible rate of heat transfer is $W_{\min}(t_{\mathrm{ent,hot}} - t_{\mathrm{ent,cold}})$, where $W_{\min}$ is the lesser of W_{hot} and W_{cold}. The rate of heat transfer in an actual heat exchanger that has an effectiveness ε is

$$q = \varepsilon W_{\min}(t_{\mathrm{ent,hot}} - t_{\mathrm{ent,cold}}) \qquad (4\text{-}61)$$

For the counterflow heat exchanger of Fig. 4-16, where

$$q = W_1(t_1 - t_2)$$

$$t_2 = t_1 - \varepsilon \frac{W_{\min}}{W_1}(t_1 - t_i) \qquad (4\text{-}62)$$

Comparing Eq. (4-62) with the alternate equation for t_2, Eq. (4-57),

$$\varepsilon = \frac{W_1}{W_{\min}}\left(\frac{1 - e^D}{W_1/W_2 - e^D}\right) \qquad (4\text{-}63)$$

Applying the effectiveness concept to a condenser such as in Fig. 4-17,

$$q = \varepsilon W_{\min}(t_c - t_1) = W_1(t_2 - t_1)$$

and since W_2 is infinite, $W_1 = W_{\min}$, so

$$t_2 = t_1 + \varepsilon(t_c - t_1)$$

which, when compared to Eq. (4-60), shows

$$\varepsilon = 1 - e^{-UA/W} \qquad (4\text{-}64)$$

For noncounterflow heat exchangers, effectiveness values are plotted in some handbooks.[1]

A comment may be made now about the influence of heat exchanger area on the effectiveness, because it will aid in a qualitative explanation of optimizations involving heat exchangers that will be performed in later chapters. Examination of both Eqs. (4-63) and (4-64) shows that an increase in Area A does not result in a proportionate increase in the effectiveness. Thus, if the area of the heat exchanger is doubled, which will roughly double the cost, the rate of heat transfer will not be doubled.

4-16 Pressure drop and pumping power A cost that appears in most economic analyses of thermal systems is the pumping cost. The size of a heat exchanger transferring heat to a liquid can be reduced, for example, if the flow rate of liquid or the velocity for a given flow rate is increased. The cost whose increase eventually overtakes the reduction in the cost of the heat exchanger as the

[1] See also W. M. Kays and A. L. London, "Compact Heat Exchangers," McGraw-Hill Book Company, New York, 1958.

velocity or flow increases is the pumping cost. Another example of the emergence of pumping cost is in the selection of optimum pipe size. The smaller the pipe, the less the first cost, but for a given flow the higher the pumping cost for the life of the system.

Since the pumping cost term appears so frequently, it is appropriate to review the expression for pumping power. The pressure drop of an incompressible fluid flowing turbulently through pipes, fittings, heat exchangers and almost any confining conduit varies as w^n

$$\Delta p = C(w)^n$$

where C is a constant, w is the mass rate of flow, and the exponent n varies between about 1.8 and 2.0. Generally the value of n is close to 2.0, except for flow in straight pipes at lower Reynolds numbers in the turbulent range.

The ideal work per unit mass required for pumping fluid in steady flow is $\int v\, dp$, and for an incompressible fluid the power required is

$$\text{Power} = \frac{w}{\rho}\,\Delta p$$

where ρ = density.

Thus the equation for the pumping power is

$$\text{Power} = \frac{C}{\rho}\,(w)^{n+1} \simeq \frac{C}{\rho}\,w^3 \tag{4-65}$$

which expression is further modified by dividing by the pump, fan, or compressor efficiency.

4-17 Turbomachinery The methods of mathematical modeling that have been explained in this chapter have been limited, generally, to expressing one variable as a function of either one or two other variables. In principle it is possible to extend these methods to functions of three variables, but the execution might be formidable. Turbomachines, such as fans, pumps, compressors, and turbines, are employed in practically all thermal systems and in these components the dependent variable may be a function of three or more independent variables. Fortunately, the tool of dimensional analysis frequently permits reducing the number of independent variables to a smaller number by treating groups of terms as individual variables. The performance of a centrifugal compressor, for example, will typically appear as in Fig. 4-19. Instead of attempting to express p_2 as a function of six variables, an equation could be

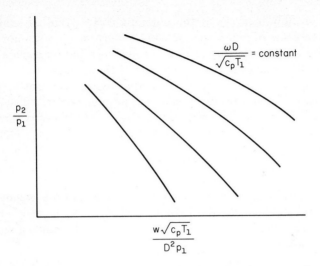

Fig. 4-19 Performance of a centrifugal compressor expressed in dimensionless groups to reduce the number of independent variables.

developed to express p_2/p_1 as a function of the two other dimensionless groups. In calculating p_2, then, the two independent variables $\omega D/\sqrt{c_p T_1}$ and $w\sqrt{c_p T_1}/(D^2 p_1)$ would be calculated first, next p_2/p_1, and, finally, p_2 would be computed from p_2/p_1 and p_1.

BIBLIOGRAPHY

Davis, D. S.: "Nomography and Empirical Equations," Reinhold Publishing Corporation, New York, 1955.

Kays, W. M., and A. L. London: "Compact Heat Exchangers," McGraw-Hill Book Company, New York, 1958.

Wambsganss, M. W., Jr.: Curve Fitting with Polynomials, *Mach. Des.*, vol. 35, no. 10, p. 167, April 25, 1963.

PROBLEMS

4-1. The power required by an automobile to overcome the air drag is to be represented by the cubic equation

$$P - P_0 = \sum_{n=1}^{3} a_n \left[\frac{3}{R} (s - s_0) \right]^n$$

where P is the power in horsepower and s is the speed in mph. The desired range of s is R and the values of P are known at s_0, s_1, s_2, and s_3, which are equally spaced in the desired range.

(a) Derive the general equation for a_1, a_2, and a_3 in terms of ΔP_1, ΔP_2, and ΔP_3, where $\Delta P_1 = P_1 - P_0$, and $\Delta P_2 = P_2 - P_0$, etc.

(b) If the following numerical values of the sP pairs apply,

$$s_0 = 40 \text{ mph} \qquad P_0 = 11.1 \text{ hp}$$
$$s_1 = 50 \text{ mph} \qquad P_1 = 20.0 \text{ hp}$$
$$s_2 = 60 \text{ mph} \qquad P_2 = 30.6 \text{ hp}$$
$$s_3 = 70 \text{ mph} \qquad P_3 = 47.5 \text{ hp}$$

calculate the numerical values of a_1, a_2, and a_3.

(c) Using the polynomial and the constants just determined, calculate the value of P at $s = 55$ mph.

Ans.: $a_3 = \frac{1}{6}(3\Delta P_1 - 3\Delta P_2 + \Delta P_3)$

$$a_2 = \frac{\Delta P_2}{2} - \Delta P_1 - 3a_3$$

$$a_1 = \Delta P_1 - a_2 - a_3$$
$$9.58, \ -1.45, \ 0.766$$
$$24.8 \text{ hp}$$

4-2. A second-degree equation of the form

$$y = a + bx + cx^2$$

has been proposed to pass through the following three (x, y) points: $(1, 3)$, $(2, 4)$, and $(2, 6)$. Proceed with the solution for a, b, and c.

(a) Describe any unusual problems encountered.

(b) Propose an alternate form of a second-degree equation that will represent these three points.

4-3. The Moody chart provides the data of Table 4-4 for the relationship of the friction factor f to the Reynolds number Re for a roughness ratio $\varepsilon/D = 0.0001$. Fit these data to an equation of the form $f = a + b(\text{Re})^m$.

Ans.: $f = 0.0113 + 1.14(\text{Re})^{-0.44}$

Table 4-4 Friction factors for Prob. 4-3

f	Re
0.0310	1×10^4
0.0214	5×10^4
0.0186	1×10^5
0.0145	5×10^5
0.0135	1×10^6
0.0123	5×10^6
0.0121	1×10^7
0.0120	5×10^7

due 2/20

4-4. Using a computer program (Gaussian elimination or any other that is available to you), solve for the x's in this set of simultaneous equations:

$$2x_1 + x_2 - 4x_3 + 6x_4 + 3x_5 - 2x_6 = 16 \qquad 1$$
$$-x_1 + 2x_2 + 3x_3 + 5x_4 - 2x_5 \qquad = -7 \qquad 2$$
$$x_1 - 2x_2 - 5x_3 + 3x_4 + 2x_5 + x_6 = 1 \qquad 3$$
$$4x_1 + 3x_2 - 2x_3 + 2x_4 \qquad + x_6 = -1 \qquad 4$$
$$3x_1 + x_2 - x_3 + 4x_4 + 3x_5 + 6x_6 = -11 \qquad 5$$
$$5x_1 + 2x_2 - 2x_3 + 3x_4 + x_5 + x_6 = 5 \qquad 6$$

Ans.: 2, -1, 1, 0, 3, -4.

4-5. The resistance to heat transfer in a hot-water, air-heating coil can be represented by the equation

$$R_t \left(\frac{\text{°F-hr}}{\text{Btu}} \right) = R_{\text{water}} + R_{\text{metal}} + R_{\text{air}} = \frac{c_1}{w^n} + R_{\text{metal}} + \frac{c_2}{G^m}$$

where w is the flow rate of water in lb/hr and G is the flow rate of air in lb/hr, and where

$$q(\text{Btu/hr}) = \frac{LMTD}{R_t}$$

Catalog data for a particular coil at various combinations of air and water flow can be used to calculate R_t. These values of R_t are shown in Table 4-5.

(a) Using the factorial plan, determine values of $R_{\text{air}} + c_3$ at the four air flows and $R_{\text{water}} + c_4$ at the four water flows.

(b) The values of c_3 and c_4 can be determined by the method of Sec. 4-10. Such a process yields values of $c_3 = 0.000152$ and $c_4 = 0.000383$. Using these values, determine n and m from a log-log graph and finally c_1 and c_2.

Ans.: $R_t = \dfrac{0.05}{w^{0.8}} + 0.00005 + \dfrac{0.056}{G^{0.55}}$

4-6. Lagrange interpolation is to be used to represent the enthalpy of saturated

Table 4-5   Heat-transfer resistance, (°F)(hr)/Btu

Water flow, lb/hr	Air-flow Rate G, lb/hr			
	5,400	10,800	16,200	21,600
1,000	0.000745	0.000588	0.000520	0.000480
2,000	0.000660	0.000503	0.000435	0.000395
4,000	0.000611	0.000454	0.000387	0.000347
6,000	0.000593	0.000436	0.000368	0.000329

air h_s as a function of the temperature t.   The pairs of values to be used as the basis are:

t, °F	h_s, Btu/lb
40	15.23
60	26.46
80	43.69
100	71.73

(a) Determine the values of the constants c_1 to c_4 in the equation, for h_s.

(b) Calculate h_s at 70°F.

Ans. to (b):   From the table of air properties $h_s = 34.09$.

4-7. Determine the values of the coefficients in the equation

$$y = a_0 + a_1x_1 + a_2x_2 + a_3x_1x_2$$

such that the equation satisfies the following sets of values:

	$x_1 = 1$	$x_1 = 2$
$x_2 = 1$	$y = 1$	$y = 2$
$x_2 = 3$	$y = 2$	$y = 5$

Ans.: ½, 0, −½, 1.

4-8. A parabola of the form

$$y = a_0 + a_1x + a_2x^2$$

is to be developed to provide the best fit (least squares) to the following pairs of points:

x	y
1	1.9
2	5.0
3	10.1
4	17.3
5	25.5

Calculate the values of a_0, a_1, a_2. For your convenience, the following summations based on the above points are provided:

$$\sum x_i = 15 \qquad \sum x_i^2 = 55 \qquad \sum x_i^3 = 225 \qquad \sum x_i^4 = 979$$

$$\sum y_i = 59.8 \qquad \sum x_i y_i = 238.9 \qquad \sum x_i^2 y_i = 1027.1$$

Ans.: 0.260, 0.678, 0.879.

4-9. The variable $z(x, y)$ is to be expressed in the equation of the form

$$z = a + bx + cy$$

The following data points are available and a least-squares fit is desired:

z	x	y
10	1	3
4	2	1
6	2	2
4	4	2

Compute the values of a, b, and c.

Ans.: 3.82, -1.18, 2.41.

4-10. To ventilate a factory building, 40,000 lb/hr of factory air at a temperature of 80°F is exhausted, and an identical flow rate of outdoor air at a tem-

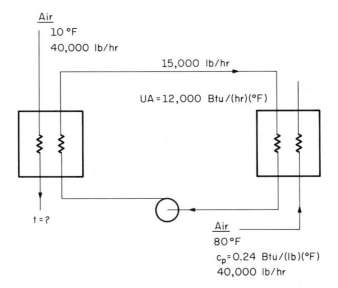

Fig. 4-20   Heat-recovery system in Prob. 4-10.

perature of 10°F is introduced to take its place.   In order to recover some of the heat of the exhaust air, heat exchangers are placed in the exhaust and ventilation-air ducts as shown in Fig. 4-20, and 15,000 lb/hr of water is pumped between the two heat exchangers.   The UA value of both of these counterflow heat exchangers is 12,000 Btu/(hr)(°F).   What is the temperature of air entering the factory?

Ans.: 36.7°F.

4-11. A double-pipe heat exchanger serves as an oil cooler with oil flowing in one direction through the inner tube and cooling water in the opposite direction through the annulus.   The oil-flow rate is 5,000 lb/hr, the oil has a specific heat of 0.40 Btu/(lb)(°F), and the water-flow rate is 4,000 lb/hr.   In a test of a prototype, oil entering at 170°F was cooled to 130°F when the entering water temperature was 90°F.   The possibility is to be considered of increasing the area of the heat exchanger by increasing the length of the double pipe.   If the flow rates, fluid properties, and entering temperatures remain unchanged, what will be the expected outlet temperature of the oil if the area is increased 20 percent?

Ans.: 125.5°F.

5
System Simulation

5-1 Introduction System simulation, as used in this chapter, is
the calculation of operating variables (pressures, temperatures,
energy, and fluid flow rates) for a system operating in a steady
state such that all energy and mass balances, all equations of state
of working substances, and the performance characteristics of all
components are satisfied.

A system is a collection of components whose performance
parameters are interrelated. The term *system simulation* means
observing a synthetic system that imitates the performance of a
real system. The type of simulation that will be studied in this
chapter will be that which can be accomplished by calculation procedures, in contrast to simulating one physical system by observing
the performance of another physical system. An example of two
corresponding physical systems is when an electrical system of
resistors and capacitors represents the heat-flow system in a solid
wall. System simulation assumes knowledge of the performance

characteristics of all components. Simulation is used when it is not possible or not economical to observe the real system. For example, it may be difficult or costly to instrument a real system in order to make detailed measurements. If the component performance is well known, however, a simulation process may pinpoint the cause of operating problems or show how the effectiveness of the system can be improved.

Another situation where system simulation is employed is where the system is still in the design stage and no real system yet exists. The performance or control of the system at off-design may be of interest, so the planned system is "run" in advance of its construction. Most thermal systems are designed for some maximum load or demand but operate most of the time at loads less than design. Economic and optimization analyses should be made throughout the range of operating conditions and not just at the design point.

This chapter first lists some of the classes of system simulation, then concentrates on just one class for the remainder of the chapter. Next are discussed the use of information-flow diagrams and the application to sequential and simultaneous calculations. The process of simulating thermal systems operating at steady state usually simmers down to the solution of simultaneous, nonlinear algebraic equations, with the procedures for their solution then being examined.

5-2 Classes of systems System simulation is a popular term and is used in different senses by various workers. We shall first list some of the classes of systems, and then designate to which type our attention will be confined.

Systems may be classified as *continuous* or *discrete*. In a continuous system, the flow through the system is that of a continuum, such as a fluid, or even solid particles flowing at such rates relative to particle sizes that the stream can be considered as a continuum. In discrete systems, the flow is treated as a certain number of integers. The analysis of the flow of people through a supermarket involving the time spent at various shopping areas and the checkout counter is a discrete system. Another example of a discrete system analysis is that performed in traffic control on turnpikes and city streets. Our concern, since it is primarily directed toward fluid and energy systems, is continuous systems.

Another classification is *deterministic* versus *stochastic*. In the deterministic analysis the input variables are precisely specified, while in the stochastic the input conditions are uncertain and are

either completely random or more commonly follow some probability distribution. In simulating the performance of a steam-
electric generating plant that supplies both process steam and
electric power to a facility, for example, a deterministic analysis
starts with one specified value of the steam demand along with one
specified value of the power demand. A stochastic analysis might
begin with some probability description of the steam and power
demands. We shall concentrate on deterministic analysis, although
certainly a series of separate deterministic analyses could be made
of different combinations of input conditions.

Finally, systems may be classed as *steady state* or *transient*.
A popular use of the term *system simulation* refers to transient analyses of systems. Transient analyses are used for such purposes as
the study of a control system in order to achieve greater precision
of control and to avoid unstable operating conditions. Essentially,
the simulation of a transient system is more difficult than the simulation of the steady-state system since the steady state falls out as
one special case of the transient analysis. This fact does not detract
from the value of steady-state simulation, however, since the entire
field of simulation has largely been limited to analysis of the smaller
and simpler systems. Pushing steady-state system analysis into
the more complex thermal system is a fruitful effort.

In summary, the simulation to be practiced here will be that
of continuous, deterministic, steady-state systems.

5-3 Information-flow diagrams Fluid- and energy-flow diagrams
are standard engineering tools. In system simulation, an equally
useful tool is the information-flow diagram. A block diagram of a
control system is an information-flow diagram wherein a block signifies that an output can be calculated when the input is known.
In contrast to the block diagram used in automatic control work
where variations with time are calculated, the steady-state diagram
implies that the parameters do not change with time. A centrifugal pump might appear in a fluid-flow diagram such as shown in
Fig. 5-1a, while in the information-flow diagram the blocks shown
in Fig. 5-1b represent functions or expressions that permit calculation of the outlet pressure for the one block and the flow rate
for the other. A block, as in Fig. 5-1b, is called a *transition
function* and may be an equation or may be tabular data to which
interpolation would be applicable.

Figure 5-1 shows only one component. To illustrate how
these individual blocks can build the information-flow diagram for
a system, consider the simple gas-turbine cycle shown in Fig. 5-2.

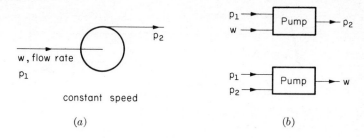

Fig. 5-1 (*a*) Centrifugal pump in fluid-flow diagram. (*b*) Possible information-flow blocks representing pump (transition functions).

The components in this cycle are the compressor, combustor, and turbine.   The turbine drives the compressor as well as provides the external power of this cycle.   The combustor receives fuel which in the burning process provides a heating rate q.   The addition of this mass of fuel results in an increase in the mass rate of flow w_3 over that leaving the compressor w_2.

The information-flow diagram is arranged in Fig. 5-3 in a manner that might be used if the shaft power E_s were to be calculated for the system with a given rate of heat input at the combustor q.   Further input information includes the ambient conditions t_1 and $p_1 = p_4$, and rotative speed.

The transition function represented by the compressor block signifies that when the speed, entering pressure, temperature, and

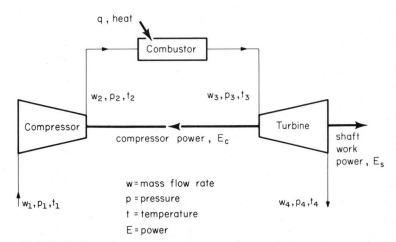

Fig. 5-2 Fluid- and energy-flow diagram of gas-turbine cycle operating at a constant speed.

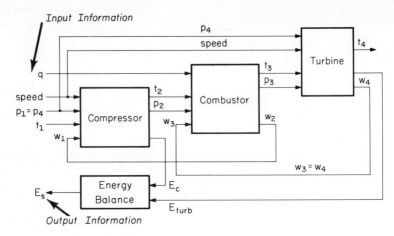

Fig. 5-3 Information-flow diagram for gas-turbine cycle. Input information: p_1, t_1, p_4, q. Output information to be calculated: E_s.

flow rate are specified, the outlet pressure and power required by
the compressor can be determined. An expected functional relationship is shown graphically in Fig. 5-4. From the graph at a
desired value of w, the values of p_2 and E_c can be read off.

The performance of a particular turbine is shown in Fig. 5-5,
where for the desired inlet temperature and pressure the power

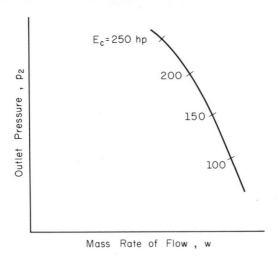

Fig. 5-4 Performance of a particular compressor
operating at a given speed and pressure, and temperature of the entering air.

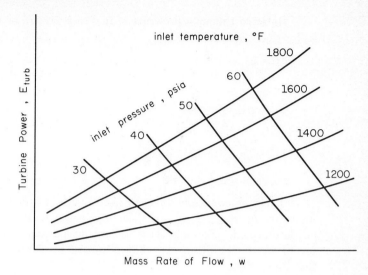

inlet temperature , °F

Fig. 5-5 Performance of a particular turbine operating at a given speed and exhaust pressure.

delivered by the turbine E_{turb} and the mass rate of flow through the turbine w can be determined. The graphs in Figs. 5-4 and 5-5 indicate that equations could be developed relating the operating parameters. The symbol ψ will be used to designate a functional relationship. For example, $\psi(x, y, z) = 0$ means that there is some equation relating x, y, and z. Under the input conditions that are imposed on the gas-turbine cycle, the following relationships are available:

Compressor:

$$\psi_1(p_2, w_1) = 0 \tag{5-1}$$

$$\psi_2(p_2, E_c) = 0 \tag{5-2}$$

From an energy balance about the compressor,

$$\psi_3(E_c, w_1, t_2, p_2) = 0 \tag{5-3}$$

Combustor:

There may be a pressure drop through the combustor,

$$\psi_4(p_2, w_1, t_2, p_3) = 0 \tag{5-4}$$

From an energy balance,

$$\psi_5(q, w_1, t_2, t_3, p_2) = 0 \tag{5-5}$$

Relating the mass flow rate of fuel to q, w_3 will be greater than w_2 by that amount

$$\psi_6(q, w_2, w_3) = 0 \tag{5-6}$$

Turbine:

$$\psi_7(t_3, p_3, w_3) = 0 \tag{5-7}$$

$$\psi_8(t_3, p_3, E_{turb}) = 0 \tag{5-8}$$

Division of turbine power,

$$\psi_9(E_{turb}, E_c, E_s) = 0 \tag{5-9}$$

The nine equations (5-1) to (5-9), contain nine unknowns, p_2, p_3, w_1, w_3, E_c, E_{turb}, E_s, t_2, and t_3, so in principle all the unknowns can be determined. Solving those operating parameters for a given set of input conditions is what is meant by simulating this gas-turbine system.

5-4 Sequential and simultaneous calculations Sometimes the arrangement of the system permits a direct numerical calculation for the first component of the system using input information. The output information for this first component is all that is needed to calculate the output information of the next component and so on to the final component of the system whose output is the output information of the system. Such a system simulation consists of *sequential* calculations. An example of a sequential calculation might occur in an on-site power-generating plant using heat recovery to generate steam for heating or refrigeration, as shown schematically in Fig. 5-6. The exhaust gas from the engine flows

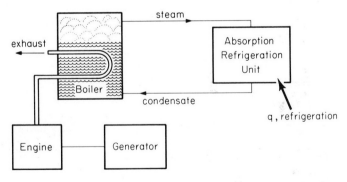

Fig. 5-6 On-site power generation with heat recovery to develop steam for refrigeration.

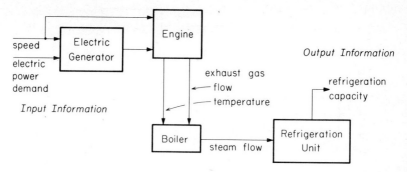

Fig. 5-7 Information-flow diagram for on-site power generating plant in Fig. 5-6.

through the boiler, which generates steam that operates an absorption refrigeration unit. If the output information is the refrigeration capacity that would be available when the unit generates a given electrical power requirement, the information-flow diagram for this calculation is shown in Fig. 5-7.

Starting with the knowledge of the engine-generator speed and electrical-power demand, and the transition functions for all of the components, the calculations may be performed in sequence through each of the components to finally arrive at the output information, which is the refrigeration capacity.

An example requiring a simultaneous calculation is shown in Fig. 5-8, where a pump with the characteristics shown in the figure delivers water to a fire-water line that contains two hydrants.

The water flow rates Q_B and Q_C through the hydrants at B and C, when open, are

$$Q_B = C \sqrt{p_B} \quad \text{and} \quad Q_C = C \sqrt{p_C}$$

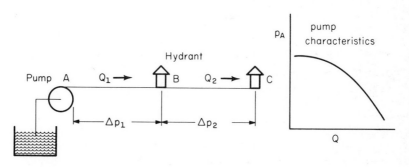

Fig. 5-8 Fire-water system and pump characteristics.

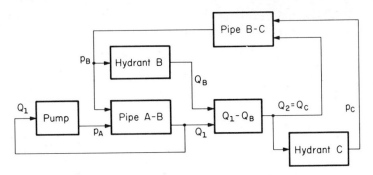

Fig. 5-9 Information-flow diagram for the fire-water system of Fig. 5-8.

where p_B and p_C are gauge pressures at positions B and C, respectively. The first glance at Fig. 5-8 may suggest that this system also simplifies to a sequential calculation. Examination of the information-flow diagram of Fig. 5-9, however, shows that there is no component where the calculation can begin and follow in sequence through the others. The transition functions for the components must be solved simultaneously. There are six components and six unknowns—three unknown flows and three pressures. In the actual solution of these equations, some of the unknowns can be eliminated easily and the number of equations correspondingly reduced.

Performing sequential calculations is straightforward. Solving a simultaneous system is a greater challenge. Since the transition functions are usually nonlinear, simulating the performance of most systems requires the solution of nonlinear, algebraic equations. The next several sections discuss the mechanics of solving such equations.

5-5 Successive substitution Probably the most straightforward method of simulating simultaneous systems is successive substitution. The successive substitution method begins with the input conditions and proceeds with the calculation until reaching a calculation block where one or more of the input variables are temporarily unknown because they are output variables of a future calculation. At such a block, assume trial values of the unknown variables and proceed through the block. After a temporarily unknown variable has been calculated, its value can be returned to where it was originally needed and be substituted for the trial value. The calculations iterate around the loop for this variable, successively substituting (which is the basis for the name of the method) until the magnitude of change during the iteration is less than a specified tolerance.

Successive substitution applied to the gas-turbine cycle in Fig. 5-3 would proceed as follows. The compressor calculations cannot be performed because w_1 is unknown, so assume a value of w_1 and calculate t_2 and p_2. Moving next to the combustor, this calculation cannot be performed because w_3 is temporarily unknown. Assuming a value of w_3, calculate $w_2 = w_1$ which can be fed back to the compressor inlet. Continue iterating about the compressor-combustor loop until the value of $w_1 = w_2$ converges. The converged value of w_1 is not a satisfactory one, however, because it was obtained with a trial value of w_3, so proceed through the turbine calculation block to compute w_4 which can be substituted successively for w_3 for iteration around the combustor-turbine loop. When w_3 and w_4 converge to a stable value, return to the compressor-combustor loop for a new iteration on w_1 and w_2. After moving back and forth between iterations of the compressor-combustor and the combustor-turbine loops until both w_1 and w_3 are stable, drop through the energy balance calculation with E_c and E_{turb} to compute E_s.

The principle of the successive substitution method is simple, and if programmed for a computer the iterations are not formidable. There is, however, a possible disadvantage and a possible pitfall to the method. The disadvantage is that equations must be available in a form where each unknown appears explicitly exactly once in the set of equations. This requirement stems from the need to be able to calculate each unknown variable by an equation somewhere in the loop.

The pitfall of the successive-substitution method is possible divergence. A simple example of both a convergent and a divergent situation is shown in the performance characteristics of the components of a fan-duct system in Fig. 5-10. Figure 5-11 shows two possible information-flow diagrams for this system. The successful simulation occurs if a pressure is assumed and the fan equation used to compute the flow rate, as in Fig. 5-11a. The flow rate thus calculated and denoted by point A in Fig. 5-10 feeds to the duct equation for determination of a new pressure. Further circuits around the loop close in on the balance point.

If, on the other hand, the information-flow diagram of Fig. 5-11b were arbitrarily chosen and a trial pressure, corresponding to point B in Fig. 5-10, were selected, calculations around the loop would result in divergence. In this example, there was only a fifty-fifty chance of convergence. On larger systems there may be

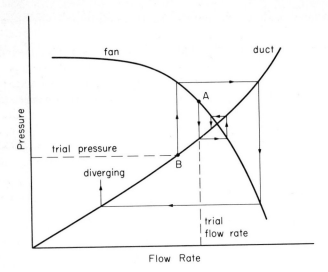

Fig. 5-10 A successful and an unsuccessful use of successive substitution for simulating a fan-duct system.

less likelihood of divergence and engineers operating simulation computer programs that use successive substitution do not report that divergence occurs frequently.   Nevertheless, the danger exists.

A method that avoids the disadvantage and pitfall of successive substitution is the simultaneous solution of the system equations using the Newton-Raphson technique. Sec. 5-9 explains this method and Secs. 5-7 and 5-8 provide the background for understanding the method.

5-6 Solution of nonlinear simultaneous equations The method for solving sets of nonlinear, simultaneous equations that will be pre-

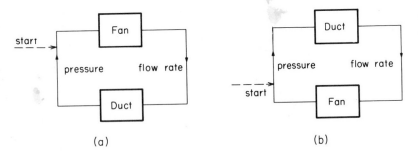

Fig. 5-11 Information flow diagram for fan-duct system of Fig. 5-10. (a) Convergent; (b) divergent.

sented in the next several sections is called the Newton-Raphson technique and is based on the Taylor series expansion. The method is an iterative one in which values for all of the unknown variables must first be assumed. The assumed values most probably will not satisfy the equations, so the solution procedure amounts to a systematic correction of the variables until the equations are satisfied within some specified tolerance.

5-7 Taylor series expansion If a function z, which is dependent upon two variables x and y, is to be expanded about the point ($x = a$ and $y = b$), the form of the series expansion is

$$z = [\text{constant}] + [\text{first-degree terms}] + [\text{second-degree terms}] + [\text{higher-degree terms}]$$

or more specifically,

$$z = [c_0] + [c_1(x - a) + c_2(y - b)] + [c_3(x - a)^2 + c_4(x - a)(y - b) + c_5(y - b)^2] + \cdots \tag{5-10}$$

Now determine the values of the constants in Eq. (5-10). If x is set equal to a and y is set equal to b, all of the terms on the right side of the equation reduce to zero except c_0, so that the value of the function at (a, b) is c_0.

$$c_0 = z(a, b) \tag{5-11}$$

To find c_1, partially differentiate Eq. (5-10) with respect to x, then set $x = a$ and $y = b$. The only term remaining on the right side of Eq. (5-10) is c_1, so

$$c_1 = \frac{\partial z\,(a,\,b)}{\partial x} \tag{5-12}$$

In a similar manner,

$$c_2 = \frac{\partial z\,(a,\,b)}{\partial y} \tag{5-13}$$

The constants c_3, c_4, and c_5 are found by partial differentiation, twice followed by substitution of $x = a$ and $y = b$ to yield the result

$$c_3 = \frac{1}{2}\frac{\partial^2 z\,(a,\,b)}{\partial x^2} \qquad c_4 = \frac{\partial^2 z\,(a,\,b)}{\partial x\,\partial y} \qquad c_5 = \frac{1}{2}\frac{\partial^2 z\,(a,\,b)}{\partial y^2} \tag{5-14}$$

For the special case where y is a function of one independent variable x, the Taylor series expansion about the point $x = a$ is

$$y = y(a) + \left[\frac{dy\,(a)}{dx}\right](x - a) + \left[\frac{1}{2}\frac{d^2 y\,(a)}{dx^2}\right](x - a)^2 + \cdots \tag{5-15}$$

The general expression for the Taylor series expansion, if y is a function of n variables $x_1, x_2, \ldots, x_n$ around the point $(x_1 = a_1, x_2 = a_2, \ldots, x_n = a_n)$ is

$$y(x_1, x_2, \ldots, x_n)$$

$$= y(a_1, a_2, \ldots, a_n) + \sum_{j=1}^{n} \left[\frac{\partial y\ (a_1, \ldots, a_n)}{\partial x_j} \right] (x_j - a_j)$$

$$+ \frac{1}{2} \sum_{i=1}^{n} \sum_{j=1}^{n} \left[\frac{\partial^2 y\ (a_1, \ldots, a_n)}{\partial x_i\ \partial x_j} \right] (x_i - a_i)(x_j - a_j) + \cdots$$

$$(5\text{-}16)$$

Example 5-1   Express the function $\ln (x^2/y)$ as a Taylor series expansion about the point $(x = 2, y = 1)$.

Solution

$$z = \ln (x^2/y) = c_0 + c_1(x - 2) + c_2(y - 1) + c_3(x - 2)^2$$
$$+ c_4(x - 2)(y - 1) + c_5(y - 1)^2 + \cdots$$

Evaluating the constants,

$$c_0 = \ln (2^2/1) = \ln 4 = 1.39$$

$$c_1 = \frac{\partial z\ (2, 1)}{\partial x} = \frac{2x/y}{x^2/y} = \frac{2}{x} = 1$$

$$c_2 = \frac{\partial z\ (2, 1)}{\partial y} = -\frac{x^2/y^2}{x^2/y} = -\frac{1}{y} = -1$$

$$c_3 = \frac{1}{2} \frac{\partial^2 z\ (2, 1)}{\partial x^2} = \frac{1}{2} \left(-\frac{2}{x^2} \right) = -\frac{1}{4}$$

$$c_4 = \frac{\partial^2 z\ (2, 1)}{\partial x\ \partial y} = 0$$

$$c_5 = \frac{1}{2} \frac{\partial^2 z\ (2, 1)}{\partial y^2} = \frac{1}{2} \left(\frac{1}{y^2} \right) = \frac{1}{2}$$

The first several terms of the expansion are, then,

$$z = 1.39 + (x - 2) - (y - 1) - \tfrac{1}{4}(x - 2)^2 + \tfrac{1}{2}(y - 1)^2 + \cdots$$

5-8   Newton-Raphson with one equation and one unknown   In the Taylor series expansion of Eq. (5-15) when x is close to a, the higher-order terms become negligible.   The equation then reduces approximately to

$$y \approx y(a) + [y'(a)](x - a) \tag{5-17}$$

Equation (5-17) is the basis of the Newton-Raphson iterative technique for solving a nonlinear algebraic equation.   Suppose that the value of x is sought that satisfies the equation

$$x + 2 = e^x \tag{5-18}$$

Define y as

$$y(x) = x + 2 - e^x \tag{5-19}$$

and, further, denote x_c as the correct value of x that solves Eq. (5-18) and makes $y = 0$

$$y(x_c) = 0 \tag{5-20}$$

The Newton-Raphson process requires an initial assumption of the value of x. Denote as x_t this trial value of x. Substituting x_t into Eq. (5-19) gives a value of y which almost certainly does not provide the desired value of $y = 0$. Specifically, if $x_t = 2$,

$$y(x_t) = x_t + 2 - e^{x_t} = 2 + 2 - 7.40 = -3.40$$

Our trial value of x is incorrect, but now the question is how the value of x should be changed in order to bring y closer to zero.

Returning to the Taylor expansion of Eq. (5-17), expand x about the unknown value of x_c

$$y(x) \approx y(x_c) + [y'(x_c)](x - x_c) \tag{5-21}$$

For $x = x_t$, Eq. (5-21) becomes

$$y(x_t) \approx y(x_c) + [y'(x_t)](x_t - x_c) \tag{5-22}$$

Equation (5-22) contains the further approximation of evaluating the derivative at x_t rather than at x_c, because the value of x_c is still unknown. From Eq. (5-20) $y(x_c) = 0$, so Eq. (5-22) can be solved approximately for the unknown value of x_c

$$x_c \approx x_t - \frac{y(x_t)}{y'(x_t)} \tag{5-23}$$

In the numerical example

$$x_c \approx 2 - \frac{-3.40}{1 - e^2} = 1.469$$

The value of $x = 1.469$ is a more correct value and should be used for the next iteration. The results of the next several iterations are

x	$y(x)$
1.469	−0.871
1.208	−0.132
1.152	−0.018

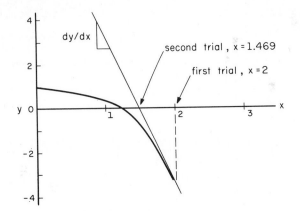

Fig. 5-12   Newton-Raphson iteration.

The graphic visualization of the iteration is shown on Fig. 5-12, where we seek the root of the equation $y = x + 2 - e^x$.   The first trial is at $x = 2$, and the deviation of y from zero divided by the slope of the curve there suggests a new trial of 1.469.

The Newton-Raphson method, while it is a powerful iteration technique, should be used carefully, because if the initial trial is too far off from the correct result, the solution may not converge. Some insight into the nature of the function being solved is, therefore, always helpful.

5-9   Sets of simultaneous equations   The solution of a nonlinear equation for the unknown variable that was discussed in Sec. 5-8 is only a special case of the solution of a set of multiple nonlinear equations.   Suppose that three nonlinear equations are to be solved for the three unknown variables x_1, x_2, and x_3.

$$f_1(x_1, x_2, x_3) = 0 \tag{5-24}$$

$$f_2(x_1, x_2, x_3) = 0 \tag{5-25}$$

$$f_3(x_1, x_2, x_3) = 0 \tag{5-26}$$

The solution procedure is an iterative one in which the following steps are followed:

Step 1   Assume trial values for the variables, denoted as $x_{1,t}$, $x_{2,t}$, $x_{3,t}$.

Step 2   Expand the function in a Taylor series about the yet unknown correct solution designated as $(x_{1,c}, x_{2,c}, x_{3,c})$.

Step 3   Solve the set of *linear* equations for improved values of the

x's and return to step 1, continuing the process until convergence has taken place to within desired limits.

In more detail, the expansion of the first of the equations,
Eq. (5-24), in a Taylor series according to Eq. (5-16) about the
point $(x_{1,c}, x_{2,c}, x_{3,c})$, including only the low-order terms is:

$$f_1(x_1, x_2, x_3) \approx f_1(x_{1,c}, x_{2,c}, x_{3,c})$$
$$+ \left[\frac{\partial f_1 (x_{1,c}, x_{2,c}, x_{3,c})}{\partial x_1} \right] (x_1 - x_{1,c})$$
$$+ \left[\frac{\partial f_1 (x_{1,c}, x_{2,c}, x_{3,c})}{\partial x_2} \right] (x_2 - x_{2,c})$$
$$+ \left[\frac{\partial f_1 (x_{1,c}, x_{2,c}, x_{3,c})}{\partial x_3} \right] (x_3 - x_{3,c}) \qquad (5\text{-}27)$$

At the trial solution point of $(x_{1,t}, x_{2,t}, x_{3,t})$, the values of the three
functions f_1, f_2, f_3 can be expressed approximately as

$$b_1 \approx a_{11}(x_{1,t} - x_{1,c}) + a_{12}(x_{2,t} - x_{2,c}) + a_{13}(x_{3,t} - x_{3,c})$$
$$(5\text{-}28)$$
$$b_2 \approx a_{21}(x_{1,t} - x_{1,c}) + a_{22}(x_{2,t} - x_{2,c}) + a_{23}(x_{3,t} - x_{3,c})$$
$$(5\text{-}29)$$
$$b_3 \approx a_{31}(x_{1,t} - x_{1,c}) + a_{32}(x_{2,t} - x_{2,c}) + a_{33}(x_{3,t} - x_{3,c})$$
$$(5\text{-}30)$$

where

$$b_k = f_k(x_{1,t}, x_{2,t}, x_{3,t}) \qquad (5\text{-}31)$$

and

$$a_{ij} = \frac{\partial f_i (x_{1,t}, x_{2,t}, x_{3,t})}{\partial x_j} \qquad (5\text{-}32)$$

Equation (5-32) shows the further approximation of evaluating
the partial derivative at the trial rather than the yet unknown correct state point.   The a and b terms are all numerical values which
when substituted into Eqs. (5-28) to (5-30) form three simultaneous
linear equations with the three unknowns: $(x_{1,t} - x_{1,c})$, $(x_{2,t} - x_{2,c})$,
and $(x_{3,t} - x_{3,c})$.   The new value of x_1 to be used, for example,
will be

$$x_{1,\text{new}} = x_{1,t} - (x_{1,t} - x_{1,c})$$

The x values are considered to be satisfactory when the f functions calculated with these x values are sufficiently close to zero or
when the corrections on all the variables are acceptably small.

Example 5-2 A water-pumping system consists of two parallel pumps drawing water from a lower reservoir and delivering it to another that is 150 ft higher, as illustrated in Fig. 5-13. In addition to overcoming the head due to the difference in elevation, the head loss in the pipes due to friction is $0.5(Q)^2$ ft of water, where Q is the total flow rate in cfs.

The head-flow characteristics of the pumps are:

Pump 1:

$$h(\text{ft}) = 250 + 30Q_1 - 6(Q_1)^2$$

Pump 2:

$$h(\text{ft}) = 300 + 20Q_2 - 12(Q_2)^2$$

where Q_1 and Q_2 are the flow rates in cfs through pumps 1 and 2, respectively. Solve for the two flow rates Q_1 and Q_2.

Solution Three equations must be satisfied—one for the piping and one for each of the two pumps. The equation for the piping is

$$h = 150 + 0.5(Q_1 + Q_2)^2$$

The three equations can be reduced to two by eliminating h.

$$250 + 30Q_1 - 6(Q_1)^2 = 150 + 0.5(Q_1 + Q_2)^2 \quad \textit{pumps in parallel}$$
$$300 + 20Q_2 - 12(Q_2)^2 = 150 + 0.5(Q_1 + Q_2)^2 \quad \textit{hpump = hpipe}$$

Next, group all of the terms on one side of the equation

$$f_1 = 100 + 30Q_1 - 6.5(Q_1)^2 - Q_1Q_2 - 0.5(Q_2)^2$$
$$f_2 = 150 + 20Q_2 - 12.5(Q_2)^2 - Q_1Q_2 - 0.5(Q_1)^2$$

The Newton-Raphson method will now be used to find the values of Q_1 and Q_2 that reduce f_1 and f_2 to zero.

First select trial values of Q_1 and Q_2; for example, $Q_1 = 6$ and $Q_2 = 4$

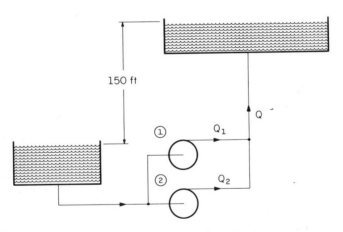

Fig. 5-13 Water-pumping system in Example 5-2.

Table 5-1 Iterations in Example 5-2

Q_1	Q_2	f_1	f_2	ΔQ_1	ΔQ_2
6.000	4.000	14.000	-12.000	-0.303	0.175
6.303	3.825	-0.545	-0.375	0.009	0.003
6.293	3.822	-0.001	0.000		

cfs. At the trial values of Q_1 and Q_2, $f_1 = 14$ and $f_2 = -12$. Next, compute the partial derivatives

$$\frac{\partial f_1}{\partial Q_1} = 30 - 13(Q_1) - Q_2 = -52$$

In a similar manner,

$$\frac{\partial f_1}{\partial Q_2} = -10 \qquad \frac{\partial f_2}{\partial Q_1} = -10 \qquad \frac{\partial f_2}{\partial Q_2} = -86$$

The partial derivatives become the coefficients in the two linear simultaneous equations:

$$-52\Delta Q_1 - 10\Delta Q_2 = 14$$
$$-10\Delta Q_1 - 86\Delta Q_2 = -12$$

where

$$\Delta Q_1 = Q_{1,\text{trial}} - Q_{1,\text{new}}$$

Solving for the ΔQ's, $\Delta Q_1 = -0.303$ and $\Delta Q_2 = 0.175$.
The improved values of Q with which to repeat the iteration are $Q_1 = 6 - (-0.303) = 6.303$ and $Q_2 = 4 - 0.175 = 3.825$.

The table of the complete set of iterations is shown in Table 5-1 resulting in the final values of $Q_1 = 6.293$ and $Q_2 = 3.822$.

5-10 Simulation of a gas-turbine cycle To provide another specific numerical illustration of system simulation, choose the gas-turbine cycle.

Example 5-3 A nonregenerative gas-turbine cycle functioning on the flow diagram shown in Fig. 5-14 operates at a constant speed of 7,200 rpm. The intake air temperature is 80°F, and the inlet and exhaust pressures are 14.7 psia.

Certain simplifications will be introduced in this solution, but it is understood that the simulation method can be extended to more refined calculations. The simplifications are:

1. Assume perfect gas properties throughout the cycle and that the c_p is constant at 0.25 Btu/(lb)(°F).
2. Neglect the mass added in the form of fuel in the combustor, the pressure drop in the combustor, and the heat transfer to the environment.

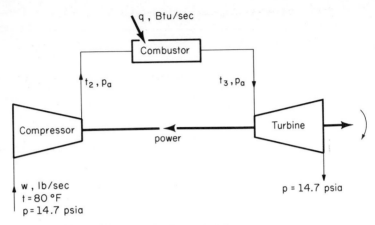

Fig. 5-14 Gas-turbine cycle in Example 5-3.

Equations for components can be fitted to curves like those shown in Figs. 5-4 and 5-5, using the technique presented in Chap. 4.

Compressor operating at 7,200 rpm

$$p_a = 48 + 3w - 0.12w^2 \tag{5-33}$$

$$E_c = 967 - 2.50p_a + 0.231p_a{}^2 \tag{5-34}$$

where p_a = high-side pressure, psia
$\quad\quad\ w$ = mass rate of flow, lb/sec
$\quad\quad\ E_c$ = power required by the compressor, Btu/sec

Turbine operating at 7,200 rpm

$$
\begin{aligned}
E_{\text{turb}} = {}&1,678 - 67.97p_a + 1.543p_a{}^2 - 3.89t_3 + 0.001111t_3{}^2 \\
&+ 0.1843p_a t_3 - 0.002161p_a{}^2 t_3 - 0.463 \times 10^{-4}p_a t_3{}^2 \\
&+ 0.617 \times 10^{-6}p_a{}^2 t_3{}^2
\end{aligned}
\tag{5-35}
$$

$$
\begin{aligned}
w = {}&16.78 + 0.487p_a + 0.00385p_a{}^2 - 0.02889t_3 \\
&+ 0.1111 \times 10^{-4}t_3{}^2 + 0.843 \times 10^{-3}p_a t_3 \\
&- 0.142 \times 10^{-4}p_a{}^2 t_3 - 0.463 \times 10^{-6}p_a t_3{}^2 \\
&+ 0.617 \times 10^{-8}p_a{}^2 t_3{}^2
\end{aligned}
\tag{5-36}
$$

where E_{turb} = power delivered by the turbine, Btu/sec
$\quad\quad\ t_3$ = turbine inlet temperature, °F

What must be the rate of heat input at the combustor in order to develop 1,400 Btu/sec shaft power? Also compute p_a, t_2, and t_3.

Solution The task of system simulation is complete when the magnitudes of the operating variables are determined such that all heat and mass balances, all equations of state, and the performance characteristics of all components are satisfied.

The necessary equations that must be satisfied will be represented as functions whose desired value is zero. The first two functions are provided by Eqs. (5-33) and (5-34).

$$f_1 = 48 + 3w - 0.12w^2 - p_a$$

and

$$f_2 = 967 - 2.5p_a + 0.231p_a{}^2 - E_c$$

Another equation from the energy balance around the compressor is that the power delivered to the compressor

$$E_c = w \text{ (increase in enthalpy)} = wc_p(t_2 - 80)$$

This equation derives from the assumption of a perfect gas in an adiabatic process. Thus,

$$f_3 = 80 + \frac{E_c}{0.25w} - t_2$$

In the combustor, the rate of heat input from the burning of fuel is q Btu/sec which equals the mass rate of flow multiplied by the increase in enthalpy.

$$q = w(\text{increase in enthalpy}) = wc_p(t_3 - t_2)$$

The function is

$$f_4 = t_2 + \frac{q}{0.25w} - t_3$$

For the turbine from Eqs. (5-35) and (5-36),

$$f_5 = 1,678 - 67.97p_a + 1.543p_a{}^2 - 3.89t_3 + 0.001111t_3{}^2$$
$$+ 0.1843p_at_3 - 0.002161p_a{}^2t_3 - 0.463 \times 10^{-4}p_at_3{}^2$$
$$+ 0.617 \times 10^{-6}p_a{}^2t_3{}^2 - E_{\text{turb}}$$

$$f_6 = 16.78 + 0.487p_a + 0.00385p_a{}^2 - 0.02889t_3$$
$$+ 0.1111 \times 10^{-4}t_3{}^2 + 0.843 \times 10^{-3}p_at_3 - 0.142 \times 10^{-4}p_a{}^2t_3$$
$$- 0.463 \times 10^{-6}p_at_3{}^2 + 0.617 \times 10^{-8}p_a{}^2t_3{}^2 - w$$

Finally the shaft power equals that developed by the turbine minus that required by the compressor.

$$f_7 = E_{\text{turb}} - E_c - 1,400$$

The solution now reduces to the process of solving the seven simultaneous, nonlinear, algebraic equations for the unknowns p_a, w, E_c, E_{turb}, t_2, t_3, and q. Examination of the equations shows that the only equation in which the variable q appears is in f_4, so this equation can be removed from the set and q solved from this equation after the other six variables are solved in the simultaneous solution. For completeness, however, we shall allow f_4 to remain in the set and permit the calculation of q along with the other unknowns.

The procedure for solving the simultaneous equations, which no doubt would be performed on a computer, is outlined in Sec. 5-9.

1. Assume trial values of the unknowns, for example,

$$p_a = 40 \text{ psia}$$
$$w = 25 \text{ lb/sec}$$
$$E_c = 1{,}500 \text{ Btu/sec}$$
$$E_{turb} = 2{,}500 \text{ Btu/sec}$$
$$t_2 = 400°\text{F}$$
$$t_3 = 1400°\text{F}$$
$$q = 6{,}000 \text{ Btu/sec}$$

2. Next calculate the values of the functions f_1 to f_7.

If they are not all close enough to zero within some prescribed tolerance the current values of the unknowns must be refined to more correct values. If the functions are satisfactorily close to zero, the solution is complete.

Certainly the functions will not be close enough to zero with the trial values. The values of the functions are, in fact, $f_1 = 8$, $f_2 = -263.4$, $f_3 = -80$, $f_4 = -40$, $f_5 = -555.3$, $f_6 = -2.8$, and $f_7 = -400$.

3. The correction process consists of solving a set of seven linear simultaneous equations, where the unknowns are the respective corrections to each variable, the coefficients are the partial derivatives, and the numerical terms on the right side of the equations are the values of the functions computed with the current values of variables.

$$\frac{\partial f_1}{\partial p_a} \Delta p_a + \frac{\partial f_1}{\partial w} \Delta w + \cdots = f_1$$

$$\cdots \cdots \cdots \cdots \cdots \cdots \cdots \cdots$$

$$\frac{\partial f_7}{\partial p_a} \Delta p_a + \cdots + \frac{\partial f_7}{\partial q} \Delta q = f_7$$

The partial derivatives are evaluated at the current values of the variables. Solution of the above simultaneous equations provides a value of Δp_a, for example, of -9.35, so the new value of p_a that will be used when looping back to step 2 will be $40 - \Delta p_a = 40 - (-9.35) = 49.35$ psia.

The results of the successive refinements of the variables until no function exceeds an absolute magnitude of 0.1 are shown in Table 5-2.

The last row of figures in Table 5-2 shows the operating variables when the system delivers a shaft power of 1,400 Btu/sec.

One of the advantages of writing a computer program to perform the above calculations is that once a program is available the operating conditions at many other input conditions can be determined simply. Also quick evaluations can be made of choosing a different component, such as a turbine or a compressor with different performance characteristics.

5-11 Utility of an information-flow diagram A system simulation certainly can be performed without resorting to an information-flow diagram, but, in many instances, the diagram has advantages. The

Table 5-2 Iterations in Example 5-3

	p_a	w	E_c	E_{turb}	t_2	t_3	q
Trial First	40.00	25.00	1,500.0	2,500.0	400.0	1,400.0	6,000.0
correction Second	49.35	24.55	1,386.0	2,786.0	301.8	1,564.3	7,890.7
correction Third	49.83	24.38	1,415.9	2,815.0	310.7	1,584.3	7,595.8
correction	49.83	24.37	1,416.0	2,816.0	312.4	1,548.4	7,532.6

first advantage of the information-flow diagram is that it aids in determining whether the simulation is sequential or simultaneous. If it is possible to start at one or more known values and proceed through the entire calculation in sequence, the simulation is sequential. When the simulation is performed by the Newton-Raphson technique, the functions can be listed and the unknowns listed without drawing a diagram. If the number of functions equals the number of unknowns and if these functions are independent, the simulation can proceed. On the other hand, the information-flow diagram is particularly useful when the successive-substitution method is used for the simulation. The successive-substitution method requires that each unknown variable appear as the output variable from a calculation block once and only once. The information-flow diagram helps in establishing this structure. In the fire-water system, for example, whose information-flow diagram is shown in Fig. 5-9, the original form of the equations for the two hydrants presented the pressure as the input and the flow rate as the output. Hydrant B is set up in that order, but the calculation for hydrant C uses the flow rate as the input and the pressure as the output in order to have a pressure available for pipe BC.

5-12 Summary Steady-state system simulation is the process of calculating the operating variables (pressure, temperatures, energy, and fluid flow rates) for a system such that all energy and mass balances, all equations of state of the working substances, and the performance characteristics of all components are satisfied. Two of the uses of system simulation are:

1. To evaluate the performance of a system operating at off-design conditions
2. To serve as a step in the optimization process

After converting the performance of the components and processes of the system to a mathematical form, the task of system simultation reduces to the solution of a set of simultaneous nonlinear equations. The two methods described in this chapter for the simultaneous solution are:

1. Successive substitution
2. The Newton-Raphson technique

BIBLIOGRAPHY

Evans, L. B., D. G. Steward, and C. R. Sprague: Computer-aided Chemical Process Design, *Chem. Eng. Progr.*, vol. 64, no. 4, pp. 39–46, 1968.
Henley, E. J., and E. M. Rosen: "Material and Energy Balance Computations," John Wiley & Sons, Inc., New York, 1969.
Mosler, H. A.: "PACER-A Digital Computer Executive Routine for Process Simulation and Design," M.S. Thesis, Purdue University, January 1964.
Naphtali, L. M.: Process Heat and Material Balances, *Chem. Eng. Progr.*, vol. 60, no. 9, pp. 70–74, September, 1964.
Stoecker, W. F.: System Simulation with the Computer, *Heating, Piping, Air Conditioning*, vol. 38, no. 3, p. 178, March, 1966.
Stoecker, W. F. (ed.): "Proposed Procedures for Simulating the Performance of Components and Systems for Energy Calculations," Am. Soc. of Htg., Ref., & A. C. Engrs., New York, 1969.

PROBLEMS

5-1. If $x \tan x = 2.0$, where x is in radians, use the Newton-Raphson method to determine the value of x.
Ans.: 1.0769.

5-2. The heat exchanger in Fig. 5-15 heats water that enters at 80°F with steam that enters as saturated vapor at 120°F and leaves as a condensate at 120°F. The flow of water is to be chosen such that the heat exchanger transfers 113,000 Btu/hr.

An increase in the rate of water flow increases the rate of heat transfer for two reasons:

1. A more favorable mean temperature difference
2. An increase in the water-side heat-transfer coefficient

Data for this heat exchanger are

$$A = 17.6 \text{ ft}^2$$

$$\frac{1}{U}\left[\frac{(\text{hr})(\text{ft}^2)(°\text{F})}{\text{Btu}}\right] = \frac{3.33}{w^{0.8}} + 0.00105$$

where w = water flow rate, lb/hr. The condenser equations from Chap. 4 apply.

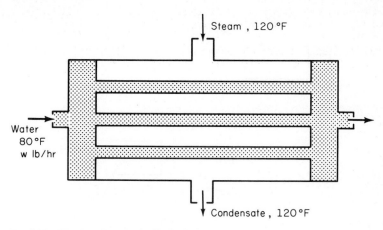

Fig. 5-15   Heat exchanger in Prob. 5-2.

By means of the Newton-Raphson method, determine the value of w that results in a heat-transfer rate of 113,000 Btu/hr.
Ans.: 5,500 lb/hr.

5-3. The operating point of a fan and duct system is to be determined. The equations for the two components are:

Duct system:

$$SP = 0.3 + (7 \times 10^{-6})Q^{1.8}$$

Fan:

$$Q = 320 - 500SP^2$$

where SP = static pressure, in. of H_2O
Q = air flow, cfs

Use the Newton-Raphson method to solve these nonlinear simultaneous equations.
Ans.: 0.425 in. of H_2O, 230 cfs.

5-4. A refrigeration plant that operates on the cycle shown in Fig. 5-16 serves as a water chiller.   Data on the individual components are as follows:

Evaporator:

$$UA = 58,000 \text{ Btu/(hr)(°F)}$$

Rate of water flow = 54,000 lb/hr

Condenser:

$$UA = 50,000 \text{ Btu/(hr)(°F)}$$

Rate of water flow = 60,000 lb/hr

The refrigeration capacity of the compressor as a function of the evaporating and condensing temperatures is given by the equation:

$$q_e(\text{Btu/hr}) = 140,000 + 21,150t_e - 65t_e^2 + 2,275t_c - 13.75t_c^2$$
$$- 231t_e t_c + 2.125t_e^2 t_c + 0.5625t_e t_c^2 - 0.00625t_e^2 t_c^2$$

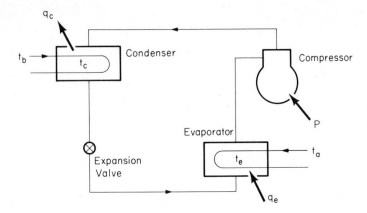

Fig. 5-16   Refrigeration plant in Prob. 5-4.

The condenser must reject q_e plus the compression power P which is represented by the equation:

$$P(\text{Btu/hr}) = 25{,}500 - 375t_e - 2.5t_e{}^2 + 712.5t_c - 3.125t_c{}^2$$
$$+ 10.375t_e t_c - 0.1375t_e{}^2 t_c + 0.04375t_e t_c{}^2$$
$$+ 0.000625\, t_e{}^2 t_c{}^2$$

Determine the condensing and evaporating temperatures q_e and P for the following combinations of inlet water temperatures:

t_a, °F	45	55	45	55
t_b, °F	80	80	100	100

Ans.: At $t_a = 45$°F and $t_b = 100$°F, $t_e = 33.77$, $t_c = 114.77$, $q_e = 399{,}260$, and $P = 101{,}657$.

100 moles/sec N_2
300 moles/sec H_2
1 mole/sec Ar

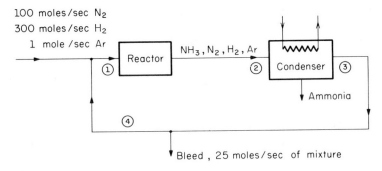

Fig. 5-17   A synthetic ammonia plant.

5-5. In a synthetic ammonia plant, as shown in Fig. 5-17, a $1:3$ mixture on a molar basis of N_2 and H_2 along with an impurity argon passes through a reactor where some of the nitrogen and hydrogen combine to form ammonia. The ammonia product that is formed leaves the system at the condenser and the remaining H_2, N_2, and Ar recycles to the reactor.

The presence of the inert gas argon is detrimental to the reaction. If no argon is present the reactor converts 60 percent of the incoming N_2 and H_2 to ammonia, but as the flow rate of argon through the reactor increases, the percent conversion decreases. The conversion efficiency follows the equation

$$\text{Conversion } (\%) = 60e^{-0.016w}$$

where w = flow rate of argon through the reactor, mol/sec.

To prevent the reaction from coming to a standstill, a continuous bleed of 25 mol/sec of mixture of N_2, H_2, and Ar is provided.

If the incoming feed consists of 100 mol/sec of N_2, 300 mol/sec of H_2, and 1 mol/sec of Ar simulate this system by successive substitution to determine the flow rate of mixture through the reactor and the rate of liquid ammonia production in mol/sec.

Ans.: 893.2 and 188 mol/sec.

6
Optimization

6-1 Introduction Optimization is the process of finding the conditions that give maximum or minimum values of a function. Optimization has always been an expected role of engineers, although sometimes on small projects the cost of engineering time may not justify an optimization effort. Often a design is difficult to optimize because of its complexity. In such cases, it may be possible to optimize subsystems and then choose the optimum combination of these subsystems. There is no assurance, however, that when using this method the true optimum is achieved.

Chapter 1 pointed out that in designing a workable system the process often consists of arbitrarily assuming certain parameters and selecting individual components around these assumptions. In contrast, when optimization is an integral part of the design, the parameters are free to float until the combination of parameters is reached which optimizes the design.

Basic to any optimization process is the decision regarding

which criterion is to be optimized. In an aircraft or space vehicle, minimum weight may be the criterion. In an automobile, the size of a system may be the criterion. Minimum cost is such a common criterion that examples are unnecessary. On the other hand, the minimum owning and operating cost, even including such factors as those studied in the chapter on economics, may not always be followed strictly. A manufacturer of domestic refrigerators, for example, does not try to design his system to provide minimum total cost to the consumer during the life of the equipment. The achievement of minimum first cost, which enhances sales, predominates over the importance of the operating cost, although the operating cost cannot be completely out of bounds. Industrial organizations often turn aside from the most economical solution by introducing human, social, and aesthetic concerns. What is happening is that their criterion function not only includes monetary factors but also includes some other factors that may, admittedly, be only vaguely defined.

Some of the topics of optimization are often practiced under the name of *operations research*. Many developments in operations research emerged from attempts to optimize mathematical models of economic systems. It is only recently that mechanical and chemical engineers have used certain of the disciplines to optimize fluid and energy-flow systems.

Component simulation and system simulation are often preliminary steps to optimizing thermal systems, since it may be necessary to simulate the performance over a wide range of operating conditions. A system that may be optimum for design loads may not be optimum over the entire range of its expected operation.

6-2 Levels of optimization Sometimes a design engineer will say, "I have optimized the design by examining four alternate concepts to do the job." He probably means that he has compared *workable* systems of four different concepts. His statement does emphasize the two levels of optimization—comparison of alternate concepts and optimization within a concept. All of the optimization methods that will be presented in the following chapters are optimizations within a concept. The flow diagram and mathematical representation of the system must be available at the beginning, and the optimization process consists of a give and take of sizes of individual components. All of this optimization is done within a given concept. There is nothing in the upcoming procedures that will kick the model over to a different one. No optimization procedure will automatically shift the system under consideration from a steam-electric generating plant to a fuel-cell concept, for example.

A complete optimization procedure, then, consists of proposing all reasonable alternate concepts, optimizing the design of each concept, and then choosing the best of the optimized designs.

6-3   Mathematical representation of optimization problems   The elements of the mathematical statement of optimization include specification of the function and the constraints.   Let y represent the function that is to be optimized, called the *objective function*, and y is a function of $x_1, x_2, \ldots, x_n$, which are called the independent variables.   The objective function, then is

$$y = y(x_1, x_2, \ldots, x_n) \rightarrow \text{optimize} \qquad (6\text{-}1)$$

In many physical situations there are constraints, some of which may be equality constraints

$$\phi_1 = \phi_1(x_1, x_2, \ldots, x_n) = 0 \qquad (6\text{-}2)$$

$$\cdots \cdots \cdots \cdots \cdots \cdots \cdots$$

$$\phi_m = \phi_m(x_1, x_2, \ldots, x_n) = 0 \qquad (6\text{-}3)$$

as well as inequality constraints

$$\psi_1 = \psi_1(x_1, x_2, \ldots, x_n) \leq L_1 \qquad (6\text{-}4)$$

$$\cdots \cdots \cdots \cdots \cdots \cdots \cdots$$

$$\psi_j = \psi_j(x_1, x_2, \ldots, x_n) \leq L_j \qquad (6\text{-}5)$$

The physical conditions dictate the sense of the inequalities in Eqs. (6-4) to (6-5).

An additive constant appearing in the objective function does not affect the values of the independent variables at which the optimum occurs.   Thus, if

$$y = a + Y(x_1, \ldots, x_n)$$

the minimum of y can be written

$$\min [a + Y(x_1, \ldots, x_n)] = a + \min [Y(x_1, \ldots, x_n)] \quad (6\text{-}6)$$

A further property of the optimum is that the maximum of a function occurs at the same state point at which the minimum of the negative of the function occurs, and

$$\max [y(x_1, \ldots, x_n)] = -\min [-y(x_1, \ldots, x_n)] \qquad (6\text{-}7)$$

6-4 A water-chilling system   A water-chilling system, shown schematically in Fig. 6-1, will be used to illustrate the mathematical statement.   The requirement of the system is that it cool 300 gpm of water from 55 to 45°F, rejecting the heat to the atmosphere through a cooling tower.   We seek a system with a minimum first cost to perform this duty.

Designate the sizes of the components in the system by x_{CP}, x_{EV}, x_{CD}, x_P, and x_{CT}, which represent the sizes of the compressor, evaporator, condenser, pump, and cooling tower, respectively.   The total cost y is the sum of the individual first plus installation costs, and this is the quantity that we wish to minimize.

$$y(x_{CP}, x_{EV}, x_{CD}, x_P, x_{CT}) \rightarrow \text{minimize} \qquad (6\text{-}8)$$

With only the statement of Eq. (6-8), the minimum could be achieved by shrinking the sizes of all components to zero.   Overlooked is the requirement that the combination of sizes be such that the water-chilling assignment can be handled.   This constraint can be expressed as in Eq. (6-9).

$$\phi(x_{CP}, x_{EV}, x_{CD}, x_P, x_{CT}) = 25{,}000 \text{ Btu/min} \qquad (6\text{-}9)$$

where ϕ is understood to mean the cooling capacity as a function of component sizes when 300 gpm of water enters at 55°F.   Actually, Eq. (6-9) could be an inequality constraint, because probably

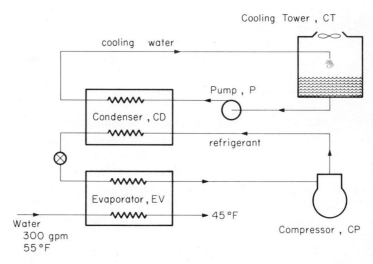

Fig. 6-1   Water-chilling unit which is being optimized for minimum first cost.

no one would object to a larger capacity than the requirement of 25,000 Btu/min.

Some practical considerations impose certain inequality constraints. The system should be designed such that the evaporating temperature t_{ev} is above 32°F or, at the lowest, 28 to 30°F to prevent water from freezing on the tube surfaces. This constraint is

$$t_{ev}(x_{CP}, x_{EV}, x_{CD}, x_P, x_{CT}) \geq 32°\text{F} \tag{6-10}$$

An extremely high discharge temperature t_d of the refrigerant leaving the compressor may impair the lubrication.

$$t_d(x_{CP}, x_{EV}, x_{CD}, x_P, x_{CT}) \leq 225°\text{F} \tag{6-11}$$

There may be other inequality constraints, such as limiting the condenser cooling-water flow in relation to the size of the cooling tower to prevent its splashing out.

The elements of the optimization problem are all present here—the objective function, equality constraints, and inequality constraints, all in terms of the independent variables, which are the sizes of components.

6-5 Optimization procedures In the next several sections of this chapter, several optimization methods will be listed. This list, while it includes most of the frequently used methods in engineering practice, is nowhere near exhaustive. In the optimization of systems, it is almost axiomatic that the objective function is dependent upon more than one variable. In fact, some thermal systems may have dozens or even hundreds of variables which demand sophisticated optimization techniques. While considerable effort may be required in the optimization process, developing mathematical relationships for the function to be optimized and the constraints may also require considerable effort.

6-6 Calculus methods—Lagrange multipliers The basis of optimization by calculus, presented in Chap. 7, is to use derivatives to indicate the optimum. The method of Lagrange multipliers performs an optimization where equality constraints exist, but the method cannot directly accommodate inequality constraints. A necessary requirement for using calculus methods is to be able to extract derivatives of the objective function and constraints.

6-7 Search methods These methods, covered in Chap. 8, involve examining a number of combinations of values of the independent variables and drawing conclusions from the magnitude of the objective function at these combinations. An obvious possibility is to calculate the value of the function at all possible combinations of, for example, 20 values distributed through the range of interest of 1 parameter, each in combination with 20 of the second variable, and so on. Such a search method is not very imaginative and is also inefficient. The search methods that are of interest are those that are efficient, particularly when applied to multivariable optimization.

When applying search methods to continuous functions, since only specific points are examined, the exact optimum can only be approached, not reached, by a finite number of trials. On the other hand, when optimizing systems where the components are available only in finite steps of sizes, search methods are often superior to calculus methods, which assume an infinite gradation of sizes.

6-8 Dynamic programming The word "programming" here and in the next several sections means "optimization" and has no direct relationship with computer programming, for example. This method of optimization, which appears in Chap. 9, is unique in that the result is an optimum *function*, rather than an optimum state point. The result of the optimization of all the other methods mentioned here is a set of values of the independent variables x_1 to x_n that result in the optimal value of the objective function y. The problem attacked by dynamic programming is one where the desired result is a *path*, for example, the best route of a gas pipeline. The result is, therefore, a function relating several variables. Dynamic programming is related to the calculus of variations, and it does in a series of discrete processes what the calculus of variations does in a continuous manner.

6-9 Geometric programming Probably the youngest of the programming family is geometric programming, which is discussed in Chap. 10. Geometric programming optimizes a function that consists of a sum of polynomials wherein the variables may appear to integer and noninteger exponents. Recalling the utility of polynomial expressions in Chap. 4, it is clear that the form of the function to which geometric programming is applicable is one that frequently occurs in thermal systems.

6-10 Linear programming Chapter 11 presents an introduction to linear programming, which is a widely used and well developed discipline that is applicable when all of the equations, Eqs. (6-1) to (6-5), are linear. The magnitude of problems now being solved by linear programming is enormous, occasionally extending into optimizations which contain several thousand variables.

6-11   Setting up the mathematical statement of an optimization problem

Example 6-1   Between two stages of air compression, the air is to be cooled from 200 to 50°F.   The facility to perform this cooling, shown in Fig. 6-2, first cools the air in a precooler and then in a refrigeration unit.   The water passes through the condenser of the refrigeration unit, then into the precooler and, finally, to a cooling tower where the heat is rejected.

The flow rate of compressed air is 2,000 scfm (standard cubic feet per minute), the standard density $= 0.075 \, \text{lb/ft}^3$, and $c_p = 0.24 \, \text{Btu/(lb)(°F)}$. The flow rate of cooling water is 18,000 lb/hr.   The system is to be designed for minimum first cost, where this first cost consists of the sum of the first costs of the refrigeration unit, precooler, and cooling tower, which are designated by x_1, x_2, and x_3, respectively, in dollars.

The expressions for the first costs are

$$x_1 = 10q_1$$

$$x_2 = \frac{25q_2}{t_4 - t_1}$$

$$x_3 = 5q_3$$

where the q's are rates of heat transfer in thousands of Btu/hr as designated in Fig. 6-2.   The water-flow rate is 18,000 lb/hr, water leaves the cooling tower at 75°F, and the power required by the refrigeration unit

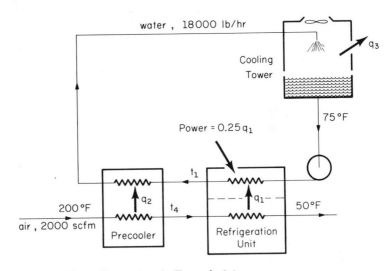

Fig. 6-2   Air-cooling system in Example 6-1.

is $0.25q_1$. Both q_1 and the thermal equivalent of the power must be removed by the condenser water.

Develop in terms of x's only: (a) the objective function and (b) the constraint equations.

Solution   The goal in this example is only to set up the problem in mathematical form, not to perform the actual optimization.   Before proceeding, however, it would be instructive to examine qualitatively the optimization features of this system.   The precooler is a simple heat exchanger so, under most operating conditions, it is less costly for a given heat-transfer rate than the refrigeration unit.   It would appear preferable, then, to do as much cooling of the air as possible with the precooler.   However, since the temperature of the water leaving the cooling tower is 75°F, it is not possible to cool air lower than 75°F in the precooler, and the final cooling of the air to 50°F must be performed by the refrigeration unit. In fact, cooling the air to 75°F in the precooler would require an infinite area, and the air temperature t_4 at optimum conditions will be higher than 75°F.   The cooling tower must reject all the heat from the system, which includes the heat from the air as well as the power to drive the refrigeration unit.   Shifting more cooling load to the refrigeration unit increases the size and cost of the cooling tower.

Writing the objective function is a simple task—the total cost is the sum of the individual costs

$$y = x_1 + x_2 + x_3$$

which is the answer to part (a).

Developing the constraint equations requires more thought.   The two essential requirements of the system are that it cool the air from 200 to 50°F and that the overall heat balance of the system be satisfied.   The rate of air flow is (2,000 cfm)(0.075 lb/ft³)(60 min/hr) = 9,000 lb/hr. The rate of heat removal from the air is (9,000 lb/hr)[(0.24 Btu/(lb)(°F)] (200 − 50°F) = 324,000 Btu/hr = 324 m-Btu/hr.

The two constraints in terms of q's are

$$q_1 + q_2 = 324 \tag{6-12}$$

and

$$q_3 - q_2 - 1.25q_1 = 0 \tag{6-13}$$

The next step is to eliminate the q's such that only the independent variables, x's, appear in the constraints.   From the expressions for the costs,

$$q_1 = 0.1x_1$$
$$q_2 = 0.04(t_4 - t_1)x_2$$
$$q_3 = 0.2x_3$$

Performing the substitutions, the constraints become

$$0.1x_1 + 0.04(t_4 - t_1)x_2 = 324 \tag{6-14}$$
$$-0.125x_1 - 0.04(t_4 - t_1)x_2 + 0.2x_3 = 0 \tag{6-15}$$

Next, the temperatures t_4 and t_1 must be replaced and this can be accomplished through heat balance relations

$$q_1 = 0.1x_1 = (9 \text{ m-lb/hr})(0.24)(t_4 - 50) = 2.16t_4 - 108$$

so

$$t_4 = 0.00463x_1 + 50$$

Another thermal quantity is the heat rejected from the refrigeration unit

$$1.25q_1 = 0.125x_1 = (18 \text{ m-lb/hr})(t_1 - 75)$$

so

$$t_1 = 75 + 0.00695x_1$$

Substituting the expressions for t_1 and t_4 into Eqs. (6-14) and (6-15) and multiplying both equations by 10 in order to achieve a slightly more convenient form, the two constraint equations become

$$x_1 + 0.0158x_1x_2 - 10x_2 = 3{,}240$$
$$1.25x_1 + 0.0158x_1x_2 - 10x_2 - 2x_3 = 0$$

which is the answer to part (b).

6-12 Discussion of Example 6-1
Optimization of thermal systems is a relatively new activity and there is much that is still to be learned. One of the gaps in knowledge is how to develop the constraint equations in a systematic manner. It is apparent in Example 6-1 that writing the objective function is a trivial task, but writing the constraints is more difficult. This is also the case in Probs. 6-1 and 6-2, at the end of the chapter, and in many thermal system optimizations.

What general conclusions can be drawn about writing the constraints? How does one know whether he has written too many? Too few? It appears that each energy flow and mass flow balance must be obeyed. Combination of constraints can be treacherous. After writing Eqs. (6-12) and (6-13), a simplification that may suggest itself is to combine the two equations and eliminate one of the q's. For example, adding the two equations results in

$$q_3 - 0.25q_1 = 324$$

which is clearly not adequate since it places no restriction on q_2, permitting it to be zero, for example. Furthermore, q_1 could reduce to zero, as long as $q_3 = 324$. The conclusion is that such a combination of constraints is invalid.

6-13 Summary While engineers have always sought to optimize their designs, it has been only within the past half dozen years— since the use of the digital computer became widespread—that sophisticated methods of optimization have become practical on complex systems. The application of optimization techniques to large-scale thermal systems is still in its infancy, but one sobering fact has emerged. Setting up the problem to the point where an optimization method can take over represents, perhaps, 70 percent of the total effort. The emphasis on optimization techniques in the next five chapters may suggest that the engineer is home free once he knows several methods. Realistically, however, the execution of the optimization can only begin when the characteristics of the physical system have been converted to the equations for the objective function and constraints.

BIBLIOGRAPHY

Beveridge, G. S. G., and R. S. Schechter: "Optimization: Theory and Practice," McGraw-Hill Book Company, New York, 1970.

Denn, M. M.: "Optimization by Variational Methods," McGraw-Hill Book Company, New York, 1969.

Rosenbrock, H. H., and C. Storey: "Computational Techniques for Chemical Engineers," Pergamon Press, New York, 1966.

Wilde, D. J., and C. S. Beightler: "Foundations of Optimization," Prentice-Hall, Inc., Englewood Cliffs, N.J., 1967.

PROBLEMS

6-1. A supersonic wind tunnel system is being designed in which the air will flow in series through a compressor, storage tank, pressure-control valve, the wind tunnel facility, and thence to exhaust. The requirement of the facility is that during tests a flow rate of 10 lb/sec must be available to the wind tunnel at a pressure of 60 psia. Heat-transfer tests are planned which require 2 min of stabilization time during which 10 lb/sec must also flow. A total of 60 min of useful test time during an 8-hour period is required, and this 60 min may be subdivided into any number of individual tests.

The mode of operation is to start the compressor and allow it to run continuously at full capacity during the 8-hour period. Between tests, the pressure in the storage tank builds up from 60 psia to the maximum of 80 psia, and during the test this stored air combines with the compressor capacity to provide the 10 lb/sec. A pressure of 60 psia is available in the storage tank at the start of the day.

The compressor and storage tank combination is to be selected for minimum total first cost. The compressor cost in dollars is given by the equation cost = 800 + 1,200S, where S is the capacity of the compressor in lb/sec. The storage tank is to be a cube for which metal costs $1.50 per ft². The mass of air that can be stored in the tank at 80 psia in excess of that at 60 psia is 0.1V lb, where V is the volume in ft³.

(a) Write the expression for the total first cost of compressor and tank in terms of S and V.

(b) Develop the constraint equation in terms of S and V to meet the operating conditions.

Ans.: $800 + 1{,}200S + 9V^{2/3}$

$$S\left(1 + \frac{0.1V}{137S + 0.0142V}\right) = 10$$

6-2. The application of a combined gas and steam-turbine plant for a liquefied petroleum gas facility at Bushton, Kansas, is described elsewhere.[1] A simplified version of this plant is shown in Fig. 6-3.

The requirements of the plant are as follows:

Power to the propane compressor	1.3×10^{10} Btu/hr
Low-pressure steam equivalent for process use	2.2×10^{10} Btu/hr
High-pressure steam equivalent for process use	3.0×10^{10} Btu/hr

In the gas-turbine plant, 20 percent of the heating value of natural gas is converted to mechanical power and the remaining 80 percent passes to the exhaust gas. As the exhaust gas flows through the boiler, 60 percent of its heat is converted to steam. The boiler is also equipped with an auxiliary burner which permits 80 percent of the heating value of the natural gas to be converted to steam. High-pressure steam flows to either process use or to the steam turbine where 15 percent of the thermal energy is converted to mechanical power.

[1] See B. F. Wobker and C. E. Knight, Mechanical Drive Combined-cycle Gas and Steam Turbines for Northern Gas Products, *ASME Paper* 67-*GT*-39, 1967.

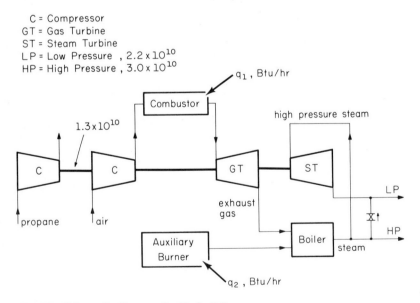

```
C = Compressor
GT = Gas Turbine
ST = Steam Turbine
LP = Low Pressure , 2.2 x 10^10
HP = High Pressure , 3.0 x 10^10
```

Fig. 6-3   Schematic diagram for Prob. 6-2.

The heat input rates at the gas turbine and the boiler are designated q_1 and q_2 Btu/hr, respectively.

The plant is to be operated in such a way that it consumes a minimum total quantity of natural gas, designated by q. In terms of q_1 and q_2

  (a) Write the objective function.

  (b) Develop the constraint equations. Acknowledge by inequalities in the constraint equations the possibility of dumping power or steam.

Ans.: $q = q_1 + q_2$ subject to:

$$q_1 + 0.441q_2 \geq 6.43 \times 10^{10}$$

$$q_1 + 1.665q_2 \geq 10.8 \times 10^{10}$$

$$q_1 + 1.175q_2 \geq 9.55 \times 10^{10}$$

7
Lagrange Multipliers

7-1 Introduction Classical methods of optimization are based on calculus and, specifically, determine the optimum value of a function as indicated by the nature of the derivatives. In order to optimize using calculus, it is necessary that the function be a differentiable equation and that any constraints be equality constraints. That there is a need for any method other than calculus, such as linear and nonlinear programming, may arise from the appearance of inequality constraints. In addition, the fact that the function is not continuous, but exists only at specific values of the parameters, rules out calculus procedures and favors such techniques as search method. On the other hand, some of the operations in the calculus methods appear in slightly revised forms in the other optimization methods, so calculus methods are important for the cases that they can solve in their own right and also to illuminate some of the procedures in noncalculus methods.

This chapter first examines unconstrained optimization, then constrained optimization, which is attacked by the method of Lagrange multipliers. Since unconstrained optimization is only a special case of constrained optimization, the method of Lagrange multipliers is applicable to all situations explored in this chapter. This chapter also explains how an optimum condition can be tested to establish whether the condition is a maximum or a minimum and, finally, introduces the concept of the sensitivity coefficient.

7-2 Unconstrained optimization If the objective function y is a function of n variables $x_1, x_2, \ldots, x_n$,

$$y = y(x_1, x_2, \ldots, x_n) \tag{7-1}$$

A critical point occurs where the derivatives are zero.

$$\frac{\partial y}{\partial x_1} = 0 \qquad \frac{\partial y}{\partial x_2} = 0 \qquad \cdots \qquad \frac{\partial y}{\partial x_n} = 0 \tag{7-2}$$

The state point where the derivatives are zero is called a critical point and it may be a maximum or minimum, which we seek, or it may be a saddle point or a ridge or valley. Further mathematical analysis may be necessary to determine the type of critical point, although in most physical situations, the nature of the point often will be obvious. We shall assume in the remainder of this chapter that Eqs. (7-2) describe a maximum or minimum.

A function y of two variables x_1 and x_2 can be represented graphically as in Fig. 7-1. The minimum exists where

$$\frac{\partial y}{\partial x_1} = 0 = \frac{\partial y}{\partial x_2}$$

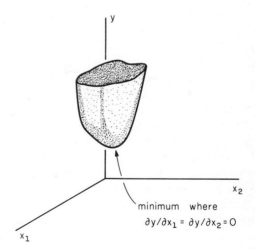

minimum where
$\partial y/\partial x_1 = \partial y/\partial x_2 = 0$

Fig. 7-1 Unconstrained optimum occurs where partial derivatives equal zero.

Example 7-1   Determine the minimum value of y and the values of x_1 and x_2 where this minimum occurs in the function

$$y = \frac{1}{x_1{}^2 x_2{}^2} + 2x_1{}^2 + 3x_2{}^3$$

Solution

$$\frac{\partial y}{\partial x_1} = -\frac{2}{x_1{}^3 x_2{}^2} + 4x_1 = 0$$

$$\frac{\partial y}{\partial x_2} = -\frac{2}{x_1{}^2 x_2{}^3} + 9x_2{}^2 = 0$$

Next, solve these two equations simultaneously.   From the first equation,

$$x_2 = \frac{1}{\sqrt{2}\,x_1{}^2}$$

Substituting into the second,

$$x_1^* = 0.972$$

Then

$$x_2^* = 0.749$$

and

$$y^* = 5.035$$

where the asterisk designates optimal values.

In Example 7-1 the two equations resulting from equating the partial derivatives to zero were such that x_1 and x_2 could be solved explicitly.   In more complex problems this may not be the case, and the iterative method for solving simultaneous nonlinear equations in Chap. 5, System Simulation, may be applied.

If the function y is dependent upon only one variable x, Eqs. (7-2) reduce to the ordinary derivative $dy/dx = 0$.

7-3   Constrained optimization   The mathematical statement of the optimization of a function under equality constraints was first given in Eqs. (6-1) to (6-3).

$$y = y(x_1, x_2, \ldots , x_n) \tag{6-1}$$

$$\phi_1(x_1, x_2, \ldots , x_n) = 0 \tag{6-2}$$

$$\cdots \cdots \cdots \cdots \cdots \cdots$$

$$\phi_m(x_1, x_2, \ldots , x_n) = 0 \tag{6-3}$$

Application of a constraint to the function shown in Fig. 7-1 would appear graphically as in Fig. 7-2.   The constraint function is

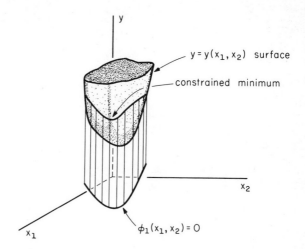

Fig. 7-2 A constrained optimum.

$\phi(x_1, x_2) = 0$, which requires that the variables x_1 and x_2 be related according to the function ϕ.

As can be seen from the figure, the constrained optimum does not necessarily occur where the partial derivatives $\partial y/\partial x_1$ and $\partial y/\partial x_2$ are zero. Some other method must be derived to perform this optimization. The method of Lagrange multipliers is one such method, but in simpler problems it may be possible to use the constraint equations to eliminate variables in the equation for y.

Example 7-2 A total of 300 lineal ft of tubes must be installed in the shell-and-tube exchanger, shown in Fig. 7-3, in order to provide the necessary heat-transfer area.

The total cost of the installation in dollars includes:

1. The cost of the tubes, which is constant at $700
2. The cost of the shell $= 25D^{2.5}L$
3. The cost of the floor space occupied by the heat exchanger $= 20DL$

The spacing of the tubes is such that 20 tubes will fit in a cross-sectional area of 1 ft^2 in the shell.

Determine the diameter and length of the heat exchanger for minimum first cost.

Solution The objective function includes the three costs,

$$\text{Cost} = 700 + 25D^{2.5}L + 20DL$$

The constraint requires the heat exchanger to include 300 ft of tubes.

$$\left[\frac{\pi D^2}{4} \text{ (ft}^2) \right] [L(\text{ft})] \, (20 \text{ tubes/ft}^2) = 300 \text{ ft}$$

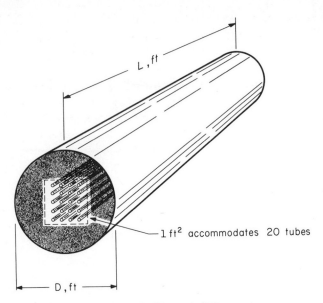

1 ft² accommodates 20 tubes

Fig. 7-3 Heat exchanger in Example 7-2.

or

$$5\pi D^2 L = 300$$

To convert this constrained optimization to an unconstrained optimization, solve for L in the constraint and substitute into the objective function.

$$\text{Cost} = 700 + \frac{1,500}{\pi} D^{0.5} + \frac{1,200}{\pi D}$$

The objective function is now in terms of D only, so differentiating and equating to zero,

$$\frac{d\,(\text{cost})}{d(D)} = \frac{750}{\pi D^{0.5}} - \frac{1,200}{\pi D^2} = 0$$

$$D^* = 1.37 \text{ ft}$$

Substituting this optimal value of D back into the constraint,

$$L^* = 10.1 \text{ ft}$$

and the minimum cost is

$$\text{Cost}^* = 700 + (25)(1.37)^{2.5}(10.1) + (20)(1.37)(10.1) = \$1,530$$

The larger the number of constraining equations, the more restricted becomes the optimization. If the number of constraint equations m equals the number of variables n no optimization is possible, since these constraint equations locate precisely one point, provided they are independent equations. Any number of constraint equations less than n offers a valid optimization.

7-4 Method of Lagrange multipliers The explanation of Lagrange multipliers in the next several sections will first present the mathematical statement, next the mechanics of performing the calculations, and finally a geometric visualization in two dimensions.

With the function to be optimized and the constraint equations as expressed in Eqs. (6-1) to (6-3), the maximum or minimum occurs where

$$\nabla y - \lambda_1 \nabla \phi_1 - \lambda_2 \nabla \phi_2 - \cdots - \lambda_m \nabla \phi_m = 0 \qquad (7-3)$$

where $\lambda_1, \lambda_2, \ldots, \lambda_m$ are called *Lagrange multipliers* and ∇ is the operator called *del* or *gradient*. The proof of Eq. (7-3) may be found in many books on advanced calculus.

Equation (7-3) is a *vector* equation in the same number of dimensions as there are variables, namely, n.

7-5 The gradient vector A *scalar* is a quantity with a magnitude but no direction, while a *vector* has both magnitude and direction. By definition, the *gradient* of a scalar is

$$\nabla y = \frac{\partial y}{\partial x_1} \bar{i}_1 + \frac{\partial y}{\partial x_2} \bar{i}_2 + \cdots + \frac{\partial y}{\partial x_n} \bar{i}_n \qquad (7-4)$$

where $\bar{i}_1, \bar{i}_2, \ldots, \bar{i}_n$ are unit vectors, which means that they have direction and their magnitudes are unity.

Suppose, for example, that a solid rectangular block has a temperature distribution that can be expressed in terms of the coordinates x_1, x_2, and x_3 as shown in Fig. 7-4.

If the temperature t is the following function of x_1, x_2, and x_3,

$$t = 2x_1 + x_1 x_2 + x_2 x_3^2$$

Then the gradient of t, ∇t, following the definition of Eq. (7-4) is

$$\nabla t = (2 + x_2)\bar{i}_1 + (x_1 + x_3^2)\bar{i}_2 + (2x_2 x_3)\bar{i}_3$$

where $\bar{i}_1$, $\bar{i}_2$, and $\bar{i}_3$ are the unit vectors in the x_1, x_2, and x_3 directions, respectively. The gradient operation is one that converts a scalar quantity to a vector quantity.

7-6 The mechanics of optimization using Lagrange multipliers Since Eq. (7-3) is a vector equation, it is really n equations because the coefficients of all of the unit vectors must sum to zero. In

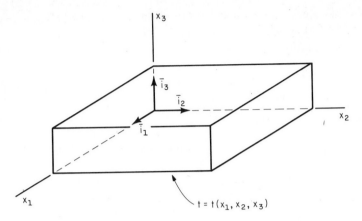

Fig. 7-4 Solid object in which a scalar, the temperature, is expressed as a function of x_1, x_2, and x_3.

addition to the n equations from Eq. (7-3), the m constraint equations are also available. Thus $n + m$ equations are available to solve for the $n + m$ unknowns $x_1 \cdots x_n$ and $\lambda_1 \cdots \lambda_m$.

Example 7-3 Solve Example 7-2 using the method of Lagrange multipliers.

Solution The objective function is

$$\text{Cost} = 700 + 25D^{2.5}L + 20DL$$

subject to the constraint:

$$5\pi D^2 L = 300$$
$$\nabla (\text{cost}) = (62.5D^{1.5}L + 20L)\bar{\imath}_1 + (25D^{2.5} + 20D)\bar{\imath}_2$$

and

$$\nabla \phi = 10\pi DL\bar{\imath}_1 + 5\pi D^2\bar{\imath}_2$$

The Lagrange multiplier equation (7-3) is

$$\nabla (\text{cost}) - \lambda \nabla \phi = 0$$

So in component form the two equations

$\bar{\imath}_1$: $62.5D^{1.5}L + 20L - \lambda 10\pi DL = 0$

$\bar{\imath}_2$: $25D^{2.5} + 20D - \lambda \pi 5D^2 = 0$

along with the constraint equation

$$5\pi D^2 L = 300$$

provide three simultaneous equations which can be solved for the unknowns D, L, and λ.

From the $\bar{\imath}_1$ equation,

$$\lambda = \frac{62.5D^{1.5} + 20}{10\pi D}$$

which, when substituted into the $\bar{\imath}_2$ equation, yields

$D^* = 1.37$ ft

Substituting this value of D into the constraint,

$L^* = 10.1$ ft

$\lambda = 2.79$

and

$\text{Cost}^* = \$1,530$

7-7 Geometric interpretation of the gradient vector
That Eq. (7-3) is the condition for optimum can be proven, although the proof will not be presented here. Instead, a geometric visualization will show that Eq. (7-3) is logical. In order to present this visualization, a geometric interpretation of the gradient vector becomes a key step. An intermediate step is to develop an expression for the vector tangent to a curve.

Suppose that a variable y is a function of x_1 and x_2,

$$y = y(x_1, x_2) \tag{7-5}$$

On the graph in Fig. 7-5, a series of curves could be developed, each

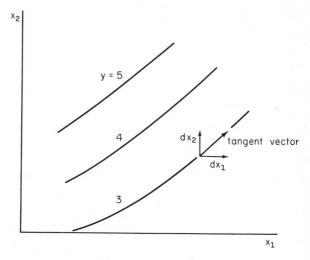

Fig. 7-5 Curves of constant y when $y = f(x_1, x_2)$.

representing a constant value of y. From calculus, we recall that

$$dy = \frac{\partial y}{\partial x_1} dx_1 + \frac{\partial y}{\partial x_2} dx_2$$

so along the line of constant y, for example, the $y = 3$ line

$$dy = 0 = \frac{\partial y}{\partial x_1} dx_1 + \frac{\partial y}{\partial x_2} dx_2$$

or

$$dx_1 = -dx_2 \frac{\partial y / \partial x_2}{\partial y / \partial x_1} \tag{7-6}$$

Any arbitrary vector on Fig. 7-5 is

$$dx_1\,\bar{\imath}_1 + dx_2\,\bar{\imath}_2$$

and any arbitrary unit vector is

$$\frac{dx_1\,\bar{\imath}_1 + dx_2\,\bar{\imath}_2}{\sqrt{(dx_1)^2 + (dx_2)^2}} \tag{7-7}$$

The special unit vector $\bar{T}$, the one that is tangent to the $y = $ constant line, has dx_1 and dx_2 related according to Eq. (7-6), so substituting dx_1 from Eq. (7-6) into Eq. (7-7) yields

$$\bar{T} = \frac{-\dfrac{\partial y / \partial x_2}{\partial y / \partial x_1}\bar{\imath}_1 + \bar{\imath}_2}{\sqrt{\left(\dfrac{\partial y / \partial x_2}{\partial y / \partial x_1}\right)^2 + 1}} = \frac{-\dfrac{\partial y}{\partial x_2}\bar{\imath}_1 + \dfrac{\partial y}{\partial x_1}\bar{\imath}_2}{\sqrt{\left(\dfrac{\partial y}{\partial x_2}\right)^2 + \left(\dfrac{\partial y}{\partial x_1}\right)^2}} \tag{7-8}$$

Equation (7-8) is the unit vector that is tangent to the $y = $ constant line.

Returning to the gradient vector, and dividing by its magnitude to obtain the unit gradient vector,

$$\frac{\nabla y}{|\nabla y|} = \frac{\dfrac{\partial y}{\partial x_1}\bar{\imath}_1 + \dfrac{\partial y}{\partial x_2}\bar{\imath}_2}{\sqrt{\left(\dfrac{\partial y}{\partial x_1}\right)^2 + \left(\dfrac{\partial y}{\partial x_2}\right)^2}} \tag{7-9}$$

The relationship between the vectors represented by Eqs. (7-8) and (7-9) is that one is perpendicular to the other. If, as in Fig. 7-6, the components of vector $\bar{Z}$ are b and c, the perpendicular vector $\bar{Z}_1$ has components c and $-b$. Thus, the components are interchanged and the sign of one is reversed.

The important conclusion reached at this point is that since the gradient vector is perpendicular to the tangent vector, *the gradient vector is normal to the line or surface of constant y.*

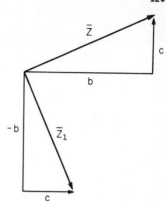

Fig. 7-6 Components of perpendicular
vectors.

A similar conclusion is applicable in three and more dimensions. In three dimensions, for example, where $y = y(x_1, x_2, x_3)$,
the curves of constant values of y become surfaces, as shown in
Fig. 7-7. In this case the gradient vector ∇y, which is

$$\nabla y = \frac{\partial y}{\partial x_1} \bar{i}_1 + \frac{\partial y}{\partial x_2} \bar{i}_2 + \frac{\partial y}{\partial x_3} \bar{i}_3 \qquad (7\text{-}10)$$

is normal to the $y =$ constant surface that passes through that
point.

The magnitude of the gradient vector indicates the rate of
change of the dependent variable with respect to the independent

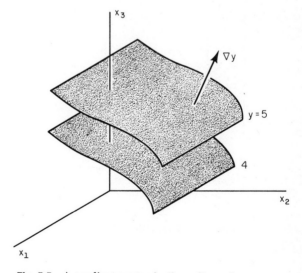

Fig. 7-7 A gradient vector in three dimensions.

variables.   Thus, if the surfaces of constant y in Fig. 7-7 are spaced widely apart, the absolute magnitude of the gradient is small.   For the time being, however, we are interested only in the fact that the gradient vector is in a *direction* normal to the constant curve or surface.

7-8   Visualization of the Lagrange multiplier method in two dimensions   The behavior of Eq. (7-3) when applied to a two-dimensional situation gives some insight into how the method of Lagrange multipliers yields an optimization.   Suppose that the minimum value of y is to be found subject to a constraint.   For this particular function, assume that the lines of constant y are as shown in Fig. 7-8.   The constraint $\phi(x_1, x_2) = 0$ is also shown on the graph.

Just from the pattern in Fig. 7-8, we immediately sense that the minimum permitted value of y exists at point A.   Point B, for example, would be rejected promptly.   If we analyze what is geometrically unique about point A, we realize that the constraint and the line of constant y are "parallel" at point A, or stated more precisely, the tangent vectors to the curves have the same direction.

A more convenient means of requiring that the tangent vectors have the same direction is to require that the normal vectors to the curves have the same direction.   Since the gradient vector is this

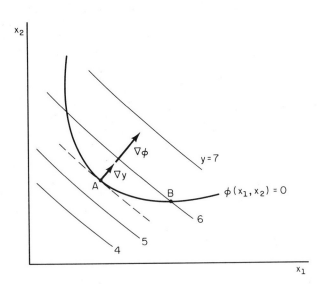

Fig. 7-8   Optimum occurs where the constraint and the lines of constant y have a common normal.

normal vector, the mathematical statement of this requirement is

$$\nabla y - \lambda \nabla \phi = 0 \tag{7-11}$$

The introduction of a constant λ is necessary because the magnitudes of ∇y and $\nabla \phi$ may be different. The sign of λ may be either plus or minus, depending upon the situation. Equation (7-11) is a vector equation, which in this two-dimensional case provides two scalar equations.

The other requirement for optimality, in addition to Eq. (7-11), is that the constraint equation be satisfied. That this is a necessity can be recognized by observing that $\nabla \phi$ is simply an equation involving x_1's and x_2's and of itself permits representation of a vector anywhere on the xy plane. We desire, however, that the vector $\nabla \phi$ be allowed only along the constraint curve, so the constraint equation in combination with Eq. (7-11) locates the optimum point.

Unconstrained optimization is a special case of constrained optimization for which the method of Lagrange multipliers also applies. If there are no constraints, $\nabla \phi = 0$ and the condition for optimality is simply $\nabla y = 0$, which is the vector form of Eq. (7-2).

7-9 Test for maximum or minimum In mathematics books that treat optimization, test procedures for determination of the nature of the "critical point" receive a major emphasis. These test procedures decide whether the point is a maximum, saddle point, ridge, or valley. These procedures are generally not so important in the physical problems, which are our concern, because the engineer usually has an insight into whether his result is a maximum or minimum. It also seems that the occurrence of saddle points, ridges, and valleys in physical problems is rare. In addition, the tests based on pure calculus usually become prohibitive when dealing with more than about three variables anyway, and we are especially interested in systems with larger numbers of components. If a test is made, then, it usually consists of testing points in the neighborhood of the optimum.

On the other hand, a brief discussion of the classical tests for maximum and minimum provide further insight into the nature of the optimization process. The discussion will be limited to the optimization of one and two variables of an unconstrained function.

Consider first the case of one independent variable $y = y(x)$ for which the minimum is sought. Suppose that the point $x = a_1$ is the position at which the minimum is expected to occur. To test whether the value of y at this point $y(a_1)$ is truly a minimum, move slightly in all possible directions from $x = a_1$ to see if a lower value of y can be found. If a lower value is available, then $y(a_1)$ is not the minimal value. For the mathematical check, expand y in a Taylor series (see Sec. 5-7) about the point $x = a_1$.

$$y(x) = y(a_1) + \frac{dy}{dx}(x - a_1) + \frac{1}{2}\frac{d^2y}{dx^2}(x - a_1)^2 + \cdots \quad (7\text{-}12)$$

First examining moves that are so small that the $(x - a_1)^2$ terms and higher-order terms can be ignored, if $dy/dx > 0$, then $y(x) > y(a_1)$ for a value of $x > a_1$, and $y(a_1)$ is still the acknowledged minimum. When x moves to a value less than a_1, however, $y(x) < y(a_1)$ and $y(a_1)$ will not be minimum. In a similar manner, it can be shown that when $dy/dx < 0$ a lower value than $y(a_1)$ can be found for $y(x)$. The only solution to the dilemma is for dy/dx to equal zero, which is the classical requirement for the optimum.

Including the influence of the next term of Eq. (7-12), $(\frac{1}{2})(d^2y/dx^2)(x - a_1)^2$, we observe that a move of x in either direction from a_1 results in a positive value of $(x - a_1)^2$, so the sign of the second derivative decides whether the optimum is a maximum or minimum.

Minimum:

$$\frac{dy}{dx} = 0 \qquad \frac{d^2y}{dx^2} > 0$$

Maximum:

$$\frac{dy}{dx} = 0 \qquad \frac{d^2y}{dx^2} < 0$$

The foregoing line of reasoning will now be extended to a function of two variables $y(x_1, x_2)$. Suppose that the expected minimum occurs at the point (a_1, a_2) and that to verify this position the point is shifted infinitesimally away from (a_1, a_2) in all possible directions. The Taylor expansion for a function of two variables is

$$\begin{aligned}
y(x_1, x_2) = {} & y(a_1, a_2) + y_1'(x_1 - a_1) + y_2'(x_2 - a_2) \\
& + \tfrac{1}{2}y_{11}''(x_1 - a_1)^2 + y_{12}''(x_1 - a_1)(x_2 - a_2) \\
& + \tfrac{1}{2}y_{22}''(x_2 - a_2)^2 + \cdots
\end{aligned} \quad (7\text{-}13)$$

where the prime on the y symbol refers to a partial differentiation with respect to the subscript.

When x_1 and x_2 move slightly off (a_1, a_2), both of the first derivatives y_1' and y_2' must be zero in order to avoid some position where $y(x_1, x_2) < y(a_1, a_2)$.

The second-order terms decide whether the optimum is a

maximum or minimum. If the combination is always positive regardless of the signs of $(x_1 - a_1)$ and $(x_2 - a_2)$, then the optimum is a minimum. If the combination is always negative, the optimum is a maximum.

A definite test for maximum and minimum is as follows. When the second derivatives are set up in matrix form and the value of the determinant is called D, where

$$D = \begin{vmatrix} y_{11}'' & y_{12}'' \\ y_{12}'' & y_{22}'' \end{vmatrix}$$

If

$$D > 0 \begin{cases} \text{If } y_{11}'' > 0 & \text{minimum} \\ \text{If } y_{11}'' < 0 & \text{maximum} \end{cases}$$

Other combinations of signs must be examined further because they may yield maxima, minima, saddle points, valleys, or ridges.

Example 7-4 Determine the optimal values of x_1 and x_2 for the function

$$y = 5x_1 - \frac{x_1{}^2 x_2}{16} + \frac{x_2{}^2}{4x_1}$$

and test whether the point is a maximum or minimum.

Solution The first derivatives are

$$\frac{\partial y}{\partial x_1} = 5 - \frac{x_1 x_2}{8} - \frac{x_2{}^2}{4x_1{}^2}$$

and

$$\frac{\partial y}{\partial x_2} = -\frac{x_1{}^2}{16} + \frac{x_2}{2x_1}$$

Equating these derivatives to zero and solving simultaneously,

$$x_1^* = 4 \quad \text{and} \quad x_2^* = 8$$

The second derivatives evaluated at x_1^* and x_2^* are

$$\frac{\partial^2 y}{\partial x_1{}^2} = 7 \qquad \frac{\partial^2 y}{\partial x_2{}^2} = \frac{1}{8} \qquad \frac{\partial^2 y}{\partial x_1\, \partial x_2} = -\frac{3}{4}$$

The determinant

$$\begin{vmatrix} 7 & -\tfrac{3}{4} \\ -\tfrac{3}{4} & \tfrac{1}{8} \end{vmatrix} > 0 \quad \text{and} \quad y_{11}'' > 0$$

so this optimal point is a *minimum*.

7-10 Sensitivity coefficients There is often an additional valuable step beyond determination of the optimal value of the objective function and the state point at which this optimum occurs. After the optimum is found subject to one or more constraints, the question that logically arises is, "What would be the effect on the optimal value of slightly relaxing the constraint?"   In a physical situation this question occurs, for example, in the analysis of how much the capacity of the system could be increased by enlarging one of the components whose performance characteristic is one of the constraint equations.

In Example 7-3 where the cost of the heat exchanger

$$\text{Cost} = 700 + 25D^{2.5}L + 20DL$$

was optimized subject to the constraint

$$5\pi D^2 L = 300$$

the question might be phrased, "What would be the increase in minimum cost, cost*, if 301 lineal ft of tubes was required rather than the original 300?"   To analyze this particular example, replace the specific value of 300 by a general symbol G and perform the optimization by the method of Lagrange multipliers.   The result would be

$$D^* = 1.37 \text{ ft}$$
$$L^* = 0.034G$$
$$\text{Cost}^* = 700 + (25)(1.37)^{2.5}(0.034G) + (20)(1.37)(0.034G)$$
$$\text{Cost}^* = 700 + 2.79G$$

We are interested in the variation of cost* with G, or more specifically, a term called the *sensitivity coefficient*, SC, which is

$$\text{SC} = \frac{\partial\,(\text{cost}^*)}{\partial G}$$

In this example, SC = 2.79, thus, at the optimal proportions a heat exchanger with an additional foot of tube would cost $2.79 more than the original one.   Referring back to the solution of Example 7-3, we note the remarkable fact that the sensitivity coefficient is precisely equal to the Lagrange multiplier λ.   This equality of SC to λ is not only true for this particular example, it is true in general. Also if there is more than one constraint, the various sensitivity coefficients are equal to the corresponding Lagrange multipliers, $\text{SC}_1 = \lambda_1, \ldots, \text{SC}_n = \lambda_n$.

The optimization process by Lagrange multipliers, therefore, offers an additional piece of useful information for possible adjustment of the physical system following preliminary optimization.

7-11 Inequality constraints The method of Lagrange multipliers applies only to the situations where the constraints are equalities and cannot be used directly with inequality constraints. This limitation should not completely rule out the employment of the method when inequalities arise, because a combination of intuition and several passes at the problem with the method may yield a solution.

As an example, suppose that the capacity of a system is stated as equal to or greater than 50,000 Btu/hr. It would be a rare case when a lower-cost system would provide 60,000 Btu/hr in comparison to the one providing 50,000 Btu/hr. This constraint, then would almost certainly be used as an equality constraint of 50,000 Btu/hr. As a further example, suppose that the temperature at some point in the process must be equal to or less than 600°F. In the first attempt at the problem, ignore the temperature constraint and, after the optimal conditions are determined, check to see if the temperature in question is above 600°F. If it is not, the constraint is not effective. If the temperature is above 600°F, rework the problem with the equality constraint of 600°F.

SEVERAL TEXTS ON ADVANCED CALCULUS FOR FURTHER STUDY OF CALCULUS METHODS OF OPTIMIZATION

Bowman, F., and F. A. Gerard: "Higher Calculus," Cambridge University Press, London, 1967.

Brand, L.: "Advanced Calculus," John Wiley & Sons, Inc., New York, 1955.

Kaplan, W.: "Advanced Calculus," Addison-Wesley Publishing Company, Inc., Reading, Mass., 1952.

Taylor, A. E.: "Advanced Calculus," Ginn and Company, Boston, 1955.

PROBLEMS

7-1. Determine the values of x_1 and x_2 that give the minimum value of y in the following equation, and also compute that minimum value.

$$y = \frac{1}{x_1} + 4x_1{}^2 x_2 - \frac{x_2{}^2}{2}$$

Ans.: $\frac{1}{2}$, 1, $\frac{5}{2}$.

7-2. A steel framework is to be constructed at a minimum cost. The cost in dollars of all the horizontal members (shown in Fig. 7-9) in one dimension is $20x_1$, and in the other horizontal direction $30x_2$. The cost in dollars of all vertical column members is $50x_3$. The frame must enclose a total volume of 900 ft³.

Solving as a constrained optimization by means of Lagrange multipliers, determine the optimal values of x_1, x_2, and x_3, and the resulting minimum cost. **Ans.:** 15, 10, and 6 ft, $900.

7-3. Determine the diameters of the circular air duct in the duct system shown schematically in Fig. 7-10 such that the static pressure at point A is a minimum.

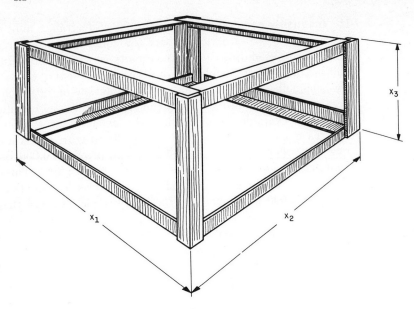

Fig. 7-9 Framework in Prob. 7-2.

Further information:
> Quantity of sheet metal available for the system is 600 ft².
> The pressure drop in a section of straight duct $= f(L/D)(V^2/2g_c)\rho$
> Assume that the friction factor $f = 0.02$.
> All outlets discharge air to atmospheric pressure.
> Neglect the influence of changes in velocity pressure.
> Neglect the pressure drop in the straightthrough section past an
> outlet.

Ans.: 1.52, 1.345, 1.02 ft.

7-4. A flow rate of 500 cfs of gas at a temperature of 120°F and a pressure of 25 psia is to be compressed to a final pressure of 2,500 psia. Influencing the type of compressor to be chosen is the characteristic of a centrifugal compressor

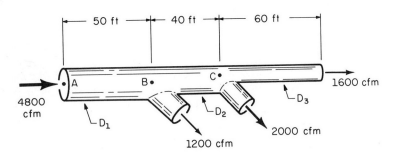

Fig. 7-10 Air duct in Prob. 7-3.

that it be capable of handling high-volume flow rates but develop only low-pressure ratios per stage. The reciprocating compressor, on the other hand, is suited to low-volume flow rates but can develop high-pressure ratios. To combine the best of two worlds, the compression will be carried out by a centrifugal compressor in series with a reciprocating compressor, as shown in Fig. 7-11. The intercooler returns the temperature of the gas to 120°F. Assume that the gas obeys perfect gas laws. The equations for the first cost of the compressors are

$$C_{cent} = 2[Q_0 \text{ (cfs)}] + 1,600 \left(\frac{p_1}{p_0}\right)$$

$$C_{recip} = 6[Q_1, \text{ (cfs)}] + 800 \left(\frac{p_2}{p_1}\right)$$

Set up the objective function for the total first cost and the constraint equation in terms of the pressure ratios, and solve for the optimal pressure ratios by the method of Lagrange multipliers.

Ans.: 7.2, 13.9.

7-5. A cylindrical oil-storage tank is to be constructed for which the following costs apply:

Metal for sides	$2.00 per ft²
Combined costs of concrete base and metal bottom	2.50 per ft²
Top	0.50 per ft²

The tank is to be constructed with dimensions such that the cost is minimum for whatever capacity is selected.

(a) One possible approach to selecting the capacity is to build the tank large enough that an additional cubic foot of capacity would cost $0.50. (Note that this does not mean $0.50 per ft³ average for the entire tank.) What is the optimal diameter and the optimal height of the tank?

Ans.: 16 ft, 12 ft.

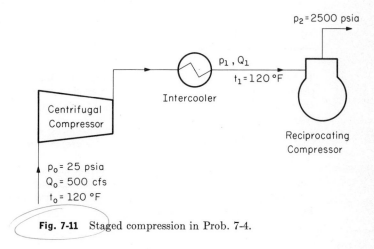

Fig. 7-11 Staged compression in Prob. 7-4.

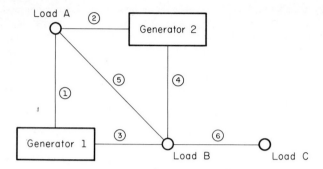

Fig. 7-12 Generating and distribution system of Prob. 7-6.

(b) Instead of the approach used in part (a), the tank is to be of such a size that the cost will be $0.60 per ft³ average for the entire storage capacity of the tank. Set up the Lagrange multiplier equations and verify that they are satisfied by an optimal diameter of 20 ft and an optimal height of 15 ft.

7-6. An electric power generating and distribution system consists of two generating plants and three loads as shown in Fig. 7-12. The loads are as follows: Load A = 40 MW, load B = 60 MW, and load C = 30 MW.

The losses in the lines are given in the following table, where the loss in line i is a function of the power carried by the line p_i:

Line losses in Prob. 7-6

Line	Loss, MW
1	$0.0010(p_1)^2$
2	$0.0012(p_2)^2$
3	$0.0007(p_3)^2$
4	$0.0006(p_4)^2$
5	$0.0008(p_5)^2$
6	$0.0011(p_6)^2$

To be precise, the line loss should be specified as a function of the power at a certain point in the line, for example, the entrance or exit, but, since the loss will be small relative to the power carried, use p_i at the point in the line most convenient for calculation.

As a first approximation in the load balances, assume that p_5 leaving load A equals p_5 entering load B, and recalculate, if necessary, after the first complete solution.

Assuming that the two generating plants are equally efficient, use the method of Lagrange multipliers to compute the optimum amount of power to be carried by each of the lines for the most efficient operation.

Ans.: 24.3, 20.3, 40, 46.4, 4.6, 31 MW.

8
Search Methods

8-1 **Introduction** In the calculus method of optimization discussed in Chap. 7, calculating the numerical value of the objective function was virtually the last step in the process. The major portion of the optimization was the determination of the values of the independent variables that resulted in the optimum. In optimization by means of search methods, an opposite procedure is followed in that values of the objective function are determined and conclusions are drawn from the values of the function at various combinations of independent variables.

Search techniques include both *elimination* and *hill-climbing* methods. Elimination methods are valuable in single-variable optimization, while hill-climbing methods are especially adaptable to functions that depend upon two or more variables. The single-variable search methods that this chapter explains are the (1) exhaustive, (2) dichotomous, (3) Fibonacci, and (4) golden-section methods. The exhaustive method is a brute-force process, while

135

the other three are more efficient methods.   Frankly speaking, it may not be worthwhile in many single-variable optimizations of thermal systems to resort to the extra effort of one of the three efficient techniques.   If the function can be calculated quickly on the computer such that a hundred or a thousand calculations do not entail excessive cost, the optimization can be performed cheaply by the exhaustive method.   On the other hand, if each calculation of the function is very lengthy, for instance, if an entire system-simulation calculation is a part of each calculation, the more efficient methods may be justified.

We shall confine our thinking to maximization in studying search methods with the understanding that a minimization problem could be converted to a maximization simply by changing the sign of the function.

8-2   Unimodal functions   A unimodal function is one in which only one peak exists in the range of interest.   Figure 8-1 shows several unimodal functions.   Not only can search methods solve the smooth function in Fig. 8-1a, but they can also solve a nondifferentiable function as in Fig. 8-1b or a discontinuous function as in Fig. 8-1c.

In optimizing nonunimodal functions where there are several peaks, the function may need to be subdivided into several parts and each part processed separately as a unimodal function.   It is understood in the future discussion of the four search methods that the function being optimized is unimodal.

8-3   Interval of uncertainty   An accepted feature of search methods is that the precise point at which the maximum occurs will never be known, and the best that can be done is to specify the *interval of uncertainty*.   The interval of uncertainty is the range of the independent variable in which the maximum is known to exist.   An

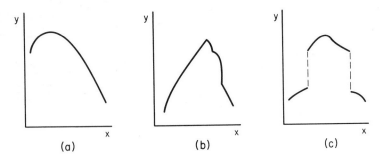

Fig. 8-1   Unimodal functions.

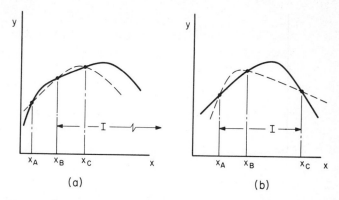

Fig. 8-2 Intervals of uncertainty I.

interval of uncertainty prevails because the search method computes the value of the function only at the discrete values of the independent variable. In Fig. 8-2a, if the values of the function at x_A, x_B, and x_C are known and are as shown in the graph, the maximum could occur between x_B and x_C as shown by the dashed line, or it could occur to the right of x_C as shown by the solid line. The most that could be concluded is that the interval of uncertainty lies to the right of x_B. When the values of the function are distributed, as in Fig. 8-2b, the maximum may occur either between x_A and x_B or between x_B and x_C. Thus, the interval of uncertainty in this case is $x_C - x_A$.

8-4 Exhaustive search Of the various search methods, the exhaustive search is the least imaginative but most widely used, and justifiably so. The method consists of calculating the value of the objective function at values of x that are spaced uniformly throughout the interval of interest. If the interval of interest, as in Fig. 8-3, is shown as I_0, and this interval is divided into eight equal intervals, assume that the values of y are calculated at the seven positions shown.

In this example, the maximum lies between x_A and x_B, so the final interval of uncertainty I is

$$I = \frac{2I_0}{8} = \frac{I_0}{4}$$

Table 8-1 lists the magnitude of the final interval of uncertainty as a function of the number of observations.

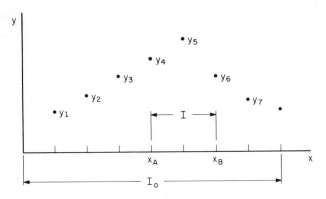

Fig. 8-3 Exhaustive search.

In general, for an exhaustive search consisting of n observations,

$$I = \frac{2I_0}{n + 1} \tag{8-1}$$

8-5 Dichotomous search The exhaustive search that was just discussed is called a *simultaneous search*, wherein all the observations are completed before any judgment is made regarding the location of the maximum.   The dichotomous search, as well as the Fibonacci and golden-section search methods, is called a *sequential search* because the results of the observations influence where future observations will be made.

In the dichotomous search, two observations that are extremely close to one another are made in the center of the interval of uncer-

Table 8-1 Final interval of uncertainty for an exhaustive search

Number of observations	I
2	$\dfrac{2I_0}{3}$
3	$\dfrac{2I_0}{4}$
4	$\dfrac{2I_0}{5}$
5	$\dfrac{2I_0}{6}$

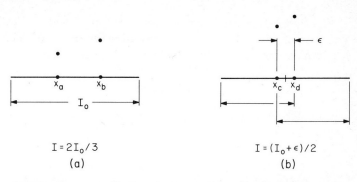

$$I = 2I_0/3$$
(a)

$$I = (I_0 + \epsilon)/2$$
(b)

Fig. 8-4 The most effective placement of two observations is as near together as possible at the center of the interval of the uncertainty.

tainty. Based on the relative values of the objective functions at those two points, almost half of the existing interval of uncertainty is eliminated. The building block of the dichotomous search is the application of a pair of observations. If points at x_a and x_b are available, as in Fig. 8-4a, the region to the left of x_a can be eliminated. The next question is, "Given a certain interval of uncertainty, where would be the most effective position to place two observations?" In Fig. 8-4, the first impulse might be to divide I_0 into thirds, but the remaining interval of uncertainty, as shown in Fig. 8-4a, would be to the right of x_a and equal to $2I_0/3$. A more effective placement of the two observations would be as close as possible to the center, yet still maintaining distinguishability of the y values. Designating as ε the distance that separates the two observations which straddle the midpoint, the final interval of uncertainty is to the right of x_c and is $(I_0 + \varepsilon)/2$.

The next pair of observations is made in a similar manner in the remaining interval of uncertainty, resulting in the interval of

Table 8-2 Final interval of uncertainty for the dichotomous search

Number of observations	I
2	$\frac{1}{2}(I_0 + \varepsilon)$
4	$\frac{1}{2}\left(\dfrac{I_0}{2} + \dfrac{\varepsilon}{2}\right) + \dfrac{\varepsilon}{2}$
6	$\frac{1}{2}\left(\dfrac{I_0}{4} + \dfrac{\varepsilon}{4} + \dfrac{\varepsilon}{2}\right) + \dfrac{\varepsilon}{2}$

uncertainty again being reduced by almost one-half with each new pair of observations. The resulting intervals of uncertainty are shown more precisely in Table 8-2. In general,

$$I = \frac{I_0}{2^{n/2}} + \varepsilon \left(1 - \frac{1}{2^{n/2}}\right)$$ (8-2)

a relation which can be proven by mathematical induction—a technique explained in Chap. 3.

8-6 Fibonacci search One of the most efficient of the single-variable search techniques is the Fibonacci method. This method was first presented by Kiefer who applied the Fibonacci number series, which was named after the thirteenth century mathematician. The technique is sequential, but the number of observations must be decided upon at the beginning of the process.

The rule for determining a Fibonacci number F is as follows:

$$F_0 = 1$$
$$F_1 = 1$$
$$F_i = F_{i-2} + F_{i-1} \quad \text{for } i \geq 2$$

Thus, after the first two Fibonacci numbers, each number is found by summing the two preceding numbers. Table 8-3 is a list of some Fibonacci numbers. These are the steps in the Fibonacci search procedure:

Step 1 Decide how many observations will be made n.

Table 8-3
Fibonacci numbers

Index i	F_i
0	1
1	1
2	2
3	3
4	5
5	8
6	13
7	21
8	34
9	55
10	89

Step 2   Place the first observations in I_0 such that the distance from one end is

$$\frac{F_{n-1}}{F_n} I_0$$

Step 3   Place the next observation in the interval of uncertainty at a position that is symmetric to the existing observation. Based on the relative values of these observations, eliminate either the region to the right of the right point or to the left of the left point.   Continue the process until one point remains. At this stage there will be one observation directly in the center of the interval of uncertainty.

Step 4   Place the last observation as close as possible to this center point and eliminate half the interval.

Example 8-1   Perform a Fibonacci search on the function $y = -(x)^2 + 4x + 2$ in the interval $0 \leq x \leq 5$.

Solution

Step 1   Arbitrarily choose $n = 4$.

Step 2   Place the first observation a distance $(F_3/F_4)I_0$ from the left end, as in Fig. 8-5.   This distance is $\frac{3}{5}$ of I_0, or $(\frac{3}{5})(5)$.   The current interval of uncertainty is $0-5$ with an observation at 3.

Step 3   The next observation symmetric in the interval of uncertainty to 3 locates this observation at $x = 2$.   Making use of the relative values of y at $x = 2$ and $x = 3$, the section $3 \leq x \leq 5$ can be eliminated.

The interval of uncertainty is now $0 \leq x \leq 3$, with the observation at $x = 2$ available.   Placing the third point symmetric to the

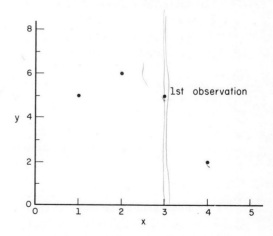

Fig. 8-5   Fibonacci search in Example 8-1.

$x = 2$ observation, locates it at $x = 1$.   The relative values of y
at $x = 1$ and $x = 2$ permit elimination of $0 \leq x < 1$.

Step 4   The final point is positioned as close as possible to the $x = 2$
point which is now located in the center of the existing interval of
uncertainty.   If this final point is placed at $x = 2 - \varepsilon$, the final
interval of uncertainty is $(2 - \varepsilon) < x \leq 3$.

The final interval of uncertainty in Example 8-1 was $I_0/5 + \varepsilon$.
In general, if n observations are employed, the final interval of
uncertainty is $I_0/F_n + \varepsilon$.

8-7   Computer procedure for a Fibonacci search   Search methods
find their greatest utility when the function is too complex to optimize by calculus or other methods.   Since the objective function is
likely to be complicated, a computer will often be used for the solution, so it is advantageous to be able to set up the Fibonacci technique in a computer program.   The possible steps in the procedure
are as follows:

1. Specify the number of trials K.
   Read in or calculate the Fibonacci numbers F(K − 1) and F(K).
   Set XL as the left border of range of interest.
   Set XR as the right border of range of interest.
2. *First point*
   X1 = XL + F(K − 1)*(XR − XL)/F(K)
   Calculate the value of the function at X1 and designate it as FX1.
   Print out X1 and FX1.
3. *Second through next-to-last point*
   Place X2 symmetrically to X1 in interval of uncertainty.
   Calculate the value of the function at X2 and designate it as FX2.
   Print out X2 and FX2.
   Test: FX1:FX2 and X1:X2.
           Based on combination of above inequalities, update XL,
           XR, and X1 and FX1.
   Loop back to the start of this section.
4. *Last point*
   Place X2 very close to X1.
   Test: FX1:FX2 and X1:X2.
           Based on the combination of the above inequalities, print
           out the interval of uncertainty and the values of the func-
           tion at each edge.

8-8   Golden-section search   A comparison of the dichotomous and
the Fibonacci methods shows that each has an advantage.   The
Fibonacci method is more efficient than the dichotomous, as will be

discussed in Sec. 8-9. On the other hand, the Fibonacci method requires an advance decision about the number of trials without any knowledge of the behavior of the function near the maximum. It may be that the function is very steep in the neighborhood of the maximum and that we would regret not having chosen a few more trials. In the dichotomous search, placing pairs of points in the middle of the interval of uncertainty can continue until either or both the interval of uncertainty and the change in value of the objective function from one trial to another is acceptably small.

The golden-section method represents a compromise, since although it is slightly less efficient than the Fibonacci method, no advance decision of the number of trials is necessary. In the Fibonacci method, the placement of the first point into the original interval of uncertainty is according to the proportion F_{k-1}/F_k. For various values of k, the ratio is as shown in Table 8-4. For large Fibonacci numbers, then, the ratio of adjacent numbers is 0.618. This value of 0.618 is the golden-section number and is used for placing the first point.

Subsequent points are placed symmetrically with respect to the already available interior point, eliminating a section after each placement. The process presumably can continue indefinitely or until either the values of y being calculated are within acceptably small changes, or the interval of uncertainty is adequately small. Actually, an anomaly arises at the fourteenth trial where the interior point lies exactly in the center of the interval of uncertainty. Some irregularity would be expected because if the value of 0.618 is applicable to the placement of the first point for all Fibonacci-type eliminations, there would be no difference between, for example,

Table 8-4   Ratio of succeeding Fibonacci numbers

k	F_{k-1}	F_k	Ratio F_{k-1}/F_k
1	1	1	1.0000
2	1	2	0.500
3	2	3	0.667
4	3	5	0.600
5	5	8	0.625
6	8	13	0.615
7	13	21	0.619
8	21	34	0.618
9	34	55	0.618
10	55	89	0.618

the search using nine trials compared to the one using ten trials. The explanation is that for large number of trials a more precise golden-section number is needed. Using more significant figures, this number is 0.618034.

8-9 Comparative effectiveness of search methods A measure of the efficiency of a search method is the reduction ratio[1] RR which is defined as the ratio of the original interval of uncertainty to the interval after n trials.

$$RR = \frac{I_0}{I_n} \tag{8-3}$$

Neglecting the penalty on the reduction ratio imposed by ε in the dichotomous and Fibonacci methods, the reduction ratios are as follows:

$$\text{Exhaustive} \quad RR = \frac{n+1}{2} \tag{8-4}$$

$$\text{Dichotomous} \quad RR = 2^{n/2} \tag{8-5}$$

$$\text{Fibonacci} \quad RR = F_n \tag{8-6}$$

The RR of the golden-section method is, for example, 250 after thirteen trials. The comparison is shown graphically in Fig. 8-6.

8-10 Multivariable searches The efficiency of such methods as Fibonacci and golden-section searches in comparison to the exhaustive search is impressive. Realistically, however, the single-variable optimizations are not the ones where high efficiencies are most

[1] See D. G. Wilde, "Optimum Seeking Methods," Prentice-Hall, Inc., Englewood Cliffs, N.J., 1964.

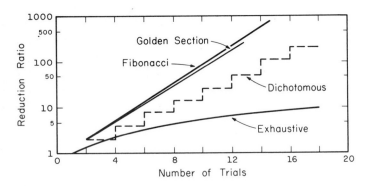

Fig. 8-6 Comparison of reduction ratios of search methods.

needed. In most single-variable optimizations encountered in thermal systems, if the search is performed on the computer the exhaustive search requires only a few more dollars of computer time and is easier to program.   Only if the calculations required for each point are extremely lengthy is there any significant advantage of one of the efficient search methods.   Cases where the calculation is extensive for a single-variable optimization are rare.

High efficiency in multivariable searches, on the other hand, may be crucial.   The optimization of thermal systems that are most significant are those involving many components and thus many variables.   Furthermore, the complexity of the equations of a many component system make the algebra extremely tedious if a calculus method is used.   This complexity breeds errors in formulation.   In many multivariable situations, the number of calculations using an exhaustive search introduces considerations of computer time.   Suppose, for example, that the optimum temperatures are sought for a seven-stage heat exchanger chain. Assume that 10 different outlet temperatures for each heat exchanger will be investigated in all combinations with the outlet temperatures of the other heat exchangers.   The total number of combinations explored will be 10^7, which, if the calculation is at all complex, makes computer time a definite concern.

There are many multivariable search methods described in the literature, but the ones to be explained in the next several sections are the lattice search, the univariate search, and the steepest-ascent (descent) method.   In the univariate search the efficient single-variable methods once again become useful.

8-11   Contour lines   When one variable is a function of two other variables, the geometric representation of the dependent variable is a surface, as shown in Fig. 8-7a.   It will be convenient in the explanation of multivariable searches in the next several sections to deal with contour curves, as in Fig. 8-7b, where the curves formed by the intersections of the surface and the constant y planes are projected down on the x_1x_2 plane.   The concepts that can be illustrated geometrically for the case of two independent variables can be extended to a larger number of variables.

8-12   Lattice search   The procedure in the lattice search is to start at one point in the region of interest and check a number of points in a grid surrounding the central point.   The surrounding point having the largest value (if a maximum is being sought) is chosen as the central point for the next search.   If no surrounding

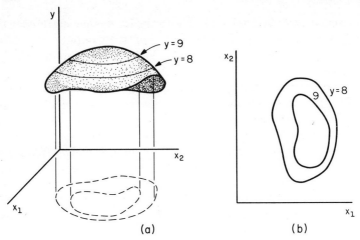

Fig. 8-7   Function of two variables represented as contour lines.

point provides a greater value of the function than the central point, the central point is the maximum.   A frequent practice is to first use a coarse grid and, after the maximum is found for that grid, subdivide the grid into smaller elements for a further search, starting from the maximum of the coarse grid.

As an example of the progression to the maximum of a function of two variables, a grid is superimposed over the contour lines of a function in Fig. 8-8.   The starting point can be selected near the

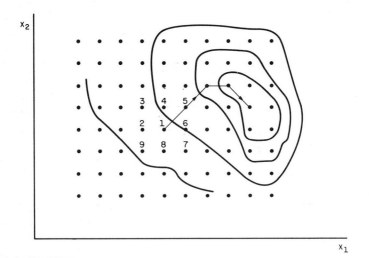

Fig. 8-8   A lattice search.

center of the region unless there is some advance knowledge of the region in which the maximum exists. Calling that first point 1, the function is evaluated at points 1 to 9. In this case, the maximum value of the function occurs at point 5, so point 5 becomes the central point for the next search.

Our special objective in applying the lattice search is to improve the efficiency compared to the exhaustive search. Since efficiency is a goal, it is attractive to consider saving information that has once been calculated and could be used again. For example, in Fig. 8-8 after moving from point 1 to point 5, the values of the function at points 1, 4, 5, and 6 are needed in deciding which way to move from point 5. Writing a computer program to accomplish this task is, however, tricky, particularly if the number of independent variables exceeds the number of dimensions possible for subscripted variables on the computer being used. It is likely, then, that the functions will be evaluated at all nine points, in the case of two variables, for each central-point location.

No definite statement can be made about the comparative efficiency of the lattice search and the exhaustive search, because the nature of the function dictates the number of trials required of the lattice search. In general, however, the ratio of the lattice trials to exhaustive trials decreases as the number of variables increases and the grid becomes finer.

8-13 Univariate search In the univariate search, the function is optimized with respect to one variable at a time. The procedure is to substitute trial values of all but one independent variable in the function and optimize the resulting function in terms of the one remaining variable. That optimal value is then substituted into the function and the function optimized with respect to another variable. The function is optimized with respect to each variable in sequence with the optimal value of a variable substituted into the function for the optimization of the succeeding variables. The process is shown graphically in Fig. 8-9 for a function of two variables. Along the line of constant x_1, which is the initial choice, the value of x_2 giving the optimal value of y is determined. This position is designated as point 1. With the value of x_2 at point 1 substituted into the function, the function is optimized with respect to x_1 which gives point 2. The process continues until the successive change of the dependent or independent variables is less than a specified tolerance.

The method chosen for performing the single-variable optimization may be by the use of calculus, where the task becomes one

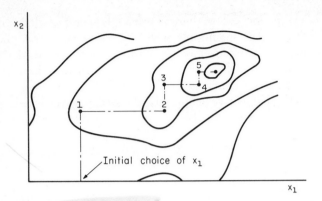

Fig. 8-9 Univariate search.

of solving one equation (usually a nonlinear one) for one variable. It is also possible to use a single-variable search—the exhaustive or one of the efficient ones, such as Fibonacci or golden section. It is this use of the efficient single-variable search that is probably the most significant application of this type of search method in thermal system optimizations.

A situation in which the univariate search can fail is when a ridge occurs in the objective function.   In the function in Fig. 8-10, for example, if a trial value of x_1 is selected as shown, the optimal value of x_2 lies on the ridge.   Substituting this optimal value of x_2, the attempt to optimize with respect to x_1 does not dislodge the point from the ridge even though the optimum has not been reached.

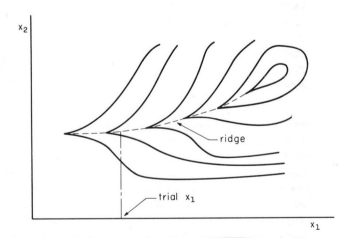

Fig. 8-10 Failure of the univariate search at a ridge.

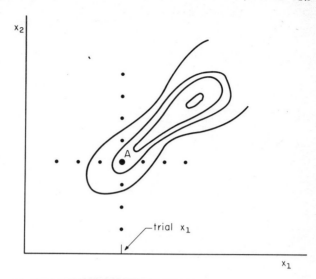

Fig. 8-11 Erroneous conclusion because of too large an interval of search.

From the purely mathematical point of view, bogging down on the ridge would appear to be a serious deficiency of the univariate search. In physical systems, however, the occurrence of ridges is rare because nature avoids discontinuities of both functions and derivatives of functions. Caution is needed, however, because even though a true ridge does not exist, problems may be encountered if the contours are very steep. If the interval chosen for the univariate search is too large, the process may stop at a nonoptimal point, such as point A in Fig. 8-11.

Another note of caution is that even though ridges might not occur in the physical system, the equations that are used to represent the physical system may accidentally contain ridges.

8-14 Steepest-ascent method As the name *steepest ascent* implies, this multivariate search method moves the state point in a direction such that the objective function changes at the greatest favorable rate. Recalling the conclusion from Sec. 7-7, the gradient vector is normal to a contour line or surface and, therefore, indicates the direction of maximum rate of change. In the function of two variables, whose contour lines are shown in Fig. 8-12, the gradient vector at point A, ∇y, is normal to the contour line at that point and indicates the direction in which y increases at the greatest rate

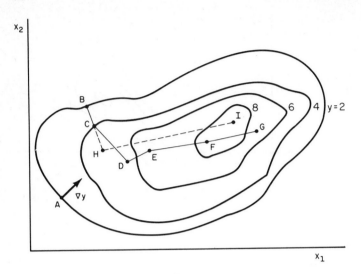

Fig. 8-12 Steepest-ascent method.

with respect to distance in the x_1x_2 plane. The equation for ∇y is

$$\nabla y = \frac{\partial y}{\partial x_1} \bar{\imath}_1 + \frac{\partial y}{\partial x_2} \bar{\imath}_2 \tag{8-7}$$

where $\bar{\imath}_1$ and $\bar{\imath}_2$ are unit vectors in the x_1 and x_2 directions, respectively.

The execution of the steepest-ascent search generally involves two separate steps:

1. Determining the direction of the move from the current location
2. Deciding upon the magnitude of the move

The first step is standard, but there are many variations of the second.

In executing step 1, the gradient equation, such as Eq. (8-7) for two independent variables, requires that the Δx_i's be related by the equation

$$\frac{\Delta x_1}{\partial y / \partial x_1} = \frac{\Delta x_2}{\partial y / \partial x_2} = \cdots = \frac{\Delta x_n}{\partial y / \partial x_n} \tag{8-8}$$

where Δx_i is the step size of the variable x_i.

Only three of the numerous methods for accomplishing step 2 will be described. The first method is to arbitrarily select a step size for one variable, Δx_1, for example, and compute $\Delta x_2 \cdots \Delta x_n$ from Eq. (8-8). This method works well until one of the partial derivatives becomes zero.

The second method is to select the step size of the independent variables such that the objective function improves by a specified amount. Designating this quantity by Δy,

$$\frac{\partial y}{\partial x_1} \Delta x_1 + \frac{\partial y}{\partial x_2} \Delta x_2 + \cdots = \Delta y \qquad (8\text{-}9)$$

Since the Δx's are related by Eq. (8-8), all the other Δx's could be expressed in terms of Δx_1 and substituted into Eq. (8-9).

$$\frac{\partial y}{\partial x_1} \Delta x_1 + \frac{\partial y}{\partial x_2} \frac{\partial y/\partial x_2}{\partial y/\partial x_1} \Delta x_1$$

$$+ \frac{\partial y}{\partial x_3} \frac{\partial y/\partial x_3}{\partial y/\partial x_1} \Delta x_1 + \cdots = \Delta y \qquad (8\text{-}10)$$

Solving for Δx_1,

$$\Delta x_1 = \frac{(\partial y/\partial x_1)\, \Delta y}{(\partial y/\partial x_1)^2 + (\partial y/\partial x_2)^2 + \cdots} \qquad (8\text{-}11)$$

Even though the Δx's are computed to provide a specified Δy, the actual Δy will differ from the desired Δy because the partial derivatives are evaluated at the starting point and do not remain constant throughout the step. In Fig. 8-12, for example, starting at point B and moving in the direction of steepest ascent with the intent of increasing y by 2.0 transfers the position to point C. The process continues to points D, E, F, and G. The move from point F to point G impaired the value of y, so a wise procedure would be to return to F and choose a smaller Δy with which to continue the search.

The third method of accomplishing step 2 is, after determining the direction indicated by the gradient vector in step 1, to proceed in that direction until reaching the optimal value in that plane. Then another gradient vector is calculated and once again the location moved in that direction until an optimal value in that plane is reached. Starting at point B in Fig. 8-12, the point moves along the dashed line to H and thence to I. The procedure for determining the optimal value in a plane in method 3 of step 2 can be converted to a single-variable search by elimination of all the x's except one by the use of Eq. (8-8). One of the efficient single-variable search methods discussed earlier in this chapter is, therefore, applicable.

The type of problem in which search methods especially would be considered are the complex ones where the manipulations of a calculus method become intractable. The entire search would be performed on a computer, probably, and for this reason even the evaluations of partial derivatives may be much easier if executed by numerical methods. The approximate value of the partial derivative at a point A using finite differences is

$$\left.\frac{\partial y}{\partial x_i}\right|_{\substack{x_{1A} \\ \vdots \\ x_{nA}}} = \frac{y[x_{1A}, x_{2A}, \ldots, (x_{iA} + \Delta), \ldots, x_{nA}] - y(x_{1A}, \ldots, x_{nA})}{\Delta}$$

Example 8-2   An insulated steel tank storing ammonia, as shown in Fig. 8-13, is equipped with a recondensation system which can control the pressure and thus the temperature of the ammonia. Two basic decisions to make in the design of this tank are the shell thickness and insulation thickness.

If the tank operates with a temperature near ambient, the pressure in the tank will be high and a thick, expensive vessel required. On the other hand, to maintain a low pressure in the tank requires more operation of the recondensation system because there will be more heat transferred from the environment unless the insulation is increased—which also adds cost.

Determine the optimum operating temperature and insulation thickness if the following costs and other data apply:

Vessel cost (\$) = $1,000 + 22(p - 14.7)^{1.2}$ for $p \geq 14.7$ psia
Insulation cost for the 600 ft^2 of heat-transfer area, (\$) = $400x^{0.9}$
Recondensation cost = \$0.01 per pound of ammonia
Lifetime hours of operation = 50,000 hr
Ambient temperature = 80°F
Average latent heat of vaporization of ammonia = 550 Btu/lb

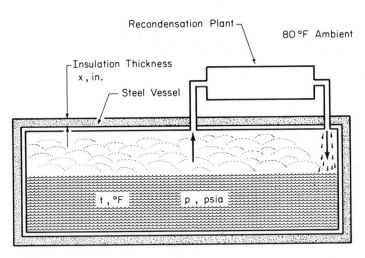

Fig. 8-13   Ammonia storage tank in Example 8-2.

Conductivity of insulation $k = 0.022$ Btu/(hr)(ft)(°F)

Pressure-temperature relation for ammonia

$$\ln p = -\frac{3{,}950}{t + 460} + 11.86$$

Solution The total lifetime cost is the sum of three individual costs: the vessel, the insulation, and the lifetime cost of recondensation. All of these costs will be expressed in terms of the operating temperature t, °F, and the insulation thickness x in.

Insulation cost:

$$IC = 400x^{0.9}$$

Vessel cost:

The saturation pressure as a function of temperature is

$$p = e^{[-3950/(t+460)+11.86]}$$

so the vessel cost VC is

$$VC = 1{,}000 + 22\{e^{[-3950/(t+460)+11.86]} - 14.7\}^{1.2}$$

Recondensation cost RC:

$$RC = [w(\text{lb/hr})](0.01 \text{ \$/lb})(50{,}000 \text{ hr})$$

where w = evaporation and recondensation rate, lb/hr
But, also

$$w = \frac{q\,(\text{Btu/hr})}{550 \text{ Btu/lb}}$$

where q = rate of heat transfer from the environment to the ammonia
Assuming that only the insulation provides any significant resistance to heat transfer,

$$q = \frac{80 - t}{x/12 \text{ (ft)}} \left(0.022 \frac{\text{Btu}}{\text{hr-ft-°F}}\right) (600 \text{ ft}^2)$$

The recondensation cost is, therefore,

$$RC = \frac{(80 - t)(0.022)(600)(0.01)(50{,}000)}{(x/12)(550)} = \frac{144(80 - t)}{x}$$

Total cost:

The total cost is the sum of the individual costs.

$$Total = IC + VC + RC$$

The search method chosen to perform this optimization will be the steepest descent, using method 3 for establishing the magnitude of the step. Arbitrarily choosing $x = 3$ in. and $t = 50$°F as the starting point, the total cost C at this position is \$5,722.56. The derivatives are

$$\frac{\partial C}{\partial x} = (400)(0.9)x^{-0.1} - \frac{(144)(80-t)}{x^2} = -157.45$$

and

$$\frac{\partial C}{\partial t} = (22)(1.2)(e^A - 14.7)^{0.2}\left[\frac{3,950}{(t+460)^2}\right]e^A - \frac{144}{x} = 4.93$$

where

$$A = -\frac{3,950}{t+460} + 11.86$$

The values of the derivatives indicate that, for a reduction in cost, x must be increased and t decreased. Furthermore, the ratio of the changes in x and t must be according to the expression $\Delta x/\Delta t = (-157.45)/(4.93)$. From the starting position, a single-variable search along the gradient finds the minimum cost to occur at $x = 3.71$ in. and $t = 49.98°F$. Table 8-5 is a summary of the calculations.

The minimum cost is \$5,339.28 when the insulation thickness is 5.94 in. and the operating temperature is 6.3°F. Both partial derivatives are approaching zero at this point. The steepest-descent calculation required an appreciable number of steps before finally homing in on the optimum. The reason is that the route passed through a curved valley, and the minimum point along the

Table 8-5   Steepest-descent search in Example 8-2

x, in.	t, °F	Cost, \$	$\partial C/\partial x$	$\partial C/\partial t$
3.00	50.00	5,722.56	−157.45	4.93
3.71	49.98	5,673.15	1.66	14.10
3.43	47.60	5,655.79	−78.35	9.03
3.86	47.55	5,639.60	0.87	13.67
3.56	42.83	5,605.88	−105.29	6.89
4.15	42.79	5,577.23	1.11	12.61
3.84	39.28	5,554.48	−82.99	7.21
4.35	39.24	5,534.40	0.55	11.57
4.05	32.96	5,497.57	−100.01	4.70
4.69	32.93	5,467.85	0.26	9.53
4.45	25.32	5,430.87	−87.53	2.89
5.07	25.30	5,405.33	−0.37	6.84
6.10	6.30	5,340.55	15.23	0.80
5.94	6.29	5,339.28	0.42	0.16

gradient alternated from one side of the valley to the other as is indicated by the alternate changing in the sign of $\partial C/\partial x$.

Further discussion of ammonia storage may be found elsewhere.[1]

8-15 Scale of the independent variables It may seem that moving in the direction of steepest ascent is the very best practice that could be hoped for. However, while in many optimization problems moving in the direction of steepest ascent is highly desirable, there are situations where it is not. In Fig. 8-14a, for example, the

[1] See N. McCloskey, Storage Facilities Associated with an Ammonia Pipeline, *ASME Paper* 69-*Pet*-21, 1969.

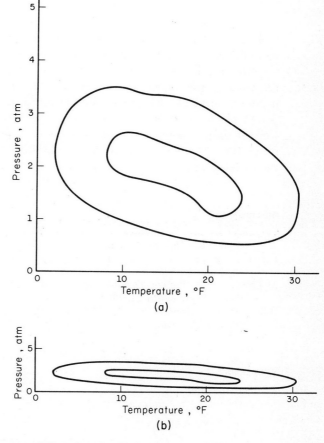

Fig. 8-14 Effect of scale of independent variables on contour lines. (a) Non-uniform scale; (b) uniform scale.

optimum for a process is a function of the pressure in atmospheres and the temperature in °F. The application of the steepest-ascent process, starting at $p = 1$ atm and $t = 25°$F, for example, would call for large moves in the pressure direction and small moves in the temperature direction. The gradient vector would not be normal to the contour lines in Fig. 8-14a because the scales of the independent variables are not the same. The elliptically shaped contours in Fig. 8-14b are very narrow, such that the steepest-ascent method practically reduces to a univariate search.

The problem of different scale may be lessened by converting one or more scales which can be accomplished in the objective function by the substitution of a new variable. In Fig. 8-14, more circular contours on uniform scales would have resulted if the pressure had been expressed in psi.

8-16 Summary Search methods have the potential of solving more difficult thermal system and other engineering-type problems, because search methods require only the capability of computing the value of the objective function at arbitrary values of the independent variables. Of the various search methods, the exhaustive search is the most easy to program for the computer and works well for simple problems. For the large-scale problems where computer time becomes a factor, more efficient search methods than the exhaustive search are demanded.

BIBLIOGRAPHY

Carnahan, B. (ed.) "Computers in Engineering Design Education," University of Michigan College of Engineering, Ann Arbor, 1966.
Kiefer, J.: Sequential Minimax Search for a Maximum, *Proc. Am. Math. Soc.*, vol. 4, p. 502, 1953.
McCloskey, N.: Storage Facilities Associated with an Ammonia Pipeline, *ASME Paper* 69-*Pet*-21, 1969.
Wilde, D. J.: "Optimum Seeking Methods," Prentice-Hall, Inc., Englewood Cliffs, N.J., 1964.

PROBLEMS

8-1. The function

$$f(x) = \frac{(\ln x) \sin \overbrace{(x^2/25)}^{\text{radians}}}{x}$$

is unimodal in the range $1 \le x \le 8$.

(a) Conduct an exhaustive search in the specified range with 0.05 increments of x, and determine the maximum value of $f(x)$ and the value of x at which it occurs.

(b) How many tests would be needed in a Fibonacci search to provide an equal or smaller level of uncertainty as the exhaustive search in part (a)?

(c) Conduct a Fibonacci search using the number of tests specified in part (b) and determine the values of x that bound the final interval of uncertainty.

Ans.: (a) 0.296, 5.95; (b) 10.

8-2. Using the golden-section search method on the original interval of uncertainty of $0 \leq x \leq 10$ for the function

$$y = 12x - 2x^2$$

carry the search for the maximum value of y until the difference between the two largest calculated values of y is 0.01 or less. What is the largest value of y and the x position at which it occurs?

Ans.: 17.997, 3.04.

8-3. An economic analysis of a proposed facility is being conducted in order to select an operating life such that the maximum uniform annual income is achieved. A short life results in high annual amortization costs, but the maintenance costs become excessive for a long life. The annual income after deducting all operating expenses, except maintenance costs, is \$180,000. The first cost of the facility is \$500,000 borrowed at 10 percent interest, compounded annually.

The maintenance costs are zero at the end of the first year, \$10,000 at the end of the second, \$20,000 at the end of the third, etc. To express these maintenance charges on an annual basis the gradient present-worth factor of Sec. 3-10 can be multiplied by the capital-recovery factor, which for the 10 percent interest is as follows:

Year	Factor	Year	Factor	Year	Factor	Year	Factor
1	0.000	6	2.224	11	4.060	16	5.552
2	0.476	7	2.622	12	4.384	17	5.801
3	0.937	8	3.008	13	4.696	18	6.058
4	1.379	9	3.376	14	5.002	19	6.295
5	1.810	10	3.730	15	5.275	20	6.500
						21	6.703

Use a Fibonacci search for integer years between 0 and 21 to find the life of the facility which results in the maximum annual profit. Omit the last calculation of the Fibonacci process since we are interested only in integer-year results.

Ans.: 12 years, \$62,760 annual income.

8-4. The exhaust gas temperature leaving a continuously operating furnace is 500°F, and a proposal is being considered to install a heat exchanger in the exhaust gas stream to generate low-pressure steam at 220°F. The question to be investigated is whether it is economical to install such a heat exchanger, and, if so, what will be its optimum size?

The following data apply:

Flow of exhaust gas	60,000 lb/hr
Specific heat of exhaust gas	0.25 Btu/(lb)(°F)
Value of the heat in the form of steam	$0.75 per million Btu
U-value of heat exchanger based on gas-side area	4 Btu/(hr)(ft²)(°F)
Cost of the heat exchanger including installation	$5 per ft² of gas-side area
Life of the installation	5 years
Interest rate	8 percent

Saturated liquid water enters the heat exchanger at 220°F and leaves as saturated vapor

Using a golden-section search and setting up a table for ease in calculation, determine the optimum heat-transfer area. Use as the original interval of uncertainty the heat-transfer area from zero to the area that results in the minimum permissible outlet gas temperature of 250°F (below which there is danger of moisture condensation in the stack). Carry on the search until the variations from one step to another in the objective function representing the annual saving is of the order of $50.

Ans.: 6,650 ft².

8-5. When an effluent, such as SO_2, discharges from a stack, as shown in Fig. 8-15, there is a certain combination of wind velocity, u mph, and distance from the stack, x ft, that results in the worst possible ground-level concentration. The ground-level concentration w can be represented by the diffusion equation

$$w = \frac{C_1}{x^{1.6}u} e^{-[C_2(h_e)^2/x^{1.6}]}$$

where C_1 and C_2 are constants and h_e is the plume height in feet. The equation for h_e is

$$h_e = h_s + \frac{k}{u}$$

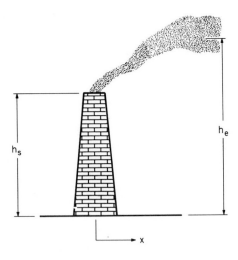

Fig. 8-15 Effluent from a stack in Prob. 8-5.

where h_s = stack height, ft

    k = a constant that is dependent upon the bouyancy of the stack gas, which is particularly influenced by its temperature

The steepest-ascent search method is to be used for finding u^* and x^* for the case where h_s = 100 ft, k = 1,000, C_1 = 5.0, and C_2 = 0.075. The trial values are x = 200 ft and u = 5 mph.

    (a) If, in the first step, the wind velocity is to be incremented by 1.0 mph, what will be the increment of change of x?

    (b) The maximum ground-level concentration in the above problem occurs when u = 10 mph and x = 150 ft. Explain why x changes so little from the trial value of 200 ft in part (a), even though it must drop by 50 ft to reach the optimum.

Ans.: (a) 0.0185 ft.

9
Dynamic Programming

9-1 Introduction Dynamic programming is a method of optimization that is applicable to either staged processes or to continuous functions that can be approximated by staged processes. The word "dynamic" has no connection with the frequent use of the word in engineering terminology where dynamic implies changes with respect to time.

Dynamic programming, as a method of optimization, is not interchangeable with such other forms of optimization as Lagrange multipliers and linear and nonlinear programming. Instead, it is related to calculus of variations whose result is an optimum *function* rather than an optimum *state point.* An optimization problem that can be subjected to dynamic programming or calculus of variations is of a different nature than the ones which submit to treatment by Lagrange multipliers and linear and nonlinear programming. Calculus of variations is used, for example, to determine the trajectory (thus, a function in spatial coordinates) that results in minimum

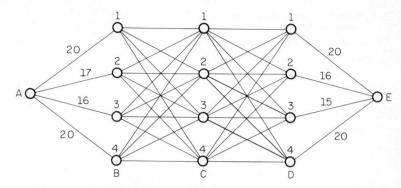

Fig. 9-1 Dynamic programming used to minimize the cost between points A and E.

fuel cost of a spacecraft. Dynamic programming can attack this same problem by dividing the total path into a number of segments and considering the continuous function as a series of steps or stages. In such an application, the finite-step approach of dynamic programming is an approximation of the calculus-of-variations method. In many engineering situations, on the other hand, the problem consists of analysis of discrete stages, such as a series of compressors, heat exchangers, or reactors. These cases fit dynamic programming exactly, and here calculus of variations would only approximate the result.

9-2 An illustration of dynamic programming An example will demonstrate some features of dynamic programming.

Example 9-1 A minimum-cost pipeline is to be constructed between points A and E, passing successively through one node of each B, C, and D as shown in Fig. 9-1. The costs from A to B and from D to E are shown in Fig. 9-1, and the costs between B and C and between C and D are given in Table 9-1.

Table 9-1   Costs from B to C and C to D

From \ To	1	2	3	4
1	12	15	24	28
2	15	16	17	21
3	24	17	16	15
4	28	21	15	12

Table 9-2   Costs from D to E

To E from	Cost	Optimum
D1	20	$\times$
D2	16	$\times$
D3	15	$\times$
D4	20	$\times$

Solution   The solution by dynamic programming begins at the destination E and works backwards, determining as the first step in the calculations the minimum-cost path between $D1$ and E, $D2$ and E, $D3$ and E, and $D4$ and E. This first stage of calculations is almost trivial since there is only one way to get, for example, from $D1$ to E. These costs are shown in Table 9-2. All of these paths are called optimum, because they represent the only possibilities of passing from the point in question to E.

Many more possibilities open up when shifting to the C points as the starting position. From point $C1$, for example, the path may pass through $D1$, $D2$, $D3$, or $D4$ on its way to E. The cost of each of these paths is computed and the minimum is noted as shown in Table 9-3.

In passing from $C1$ to E, the path of minimum cost is through $D2$.

Table 9-3   Costs from C to E

To E from	Through	Cost C to D	D to E	Total	Optimum
C1	D1	12	20	32	
C1	D2	15	16	31	$\times$
C1	D3	24	15	39	
C1	D4	28	20	48	
C2	D1	15	20	35	
C2	D2	16	16	32	$\times$ or
C2	D3	17	15	32	$\times$
C2	D4	21	20	41	
C3	D1	24	20	44	
C3	D2	17	16	33	
C3	D3	16	15	31	$\times$
C3	D4	15	20	35	
C4	D1	28	20	48	
C4	D2	21	16	37	
C4	D3	15	15	30	$\times$
C4	D4	12	20	32	

Table 9-4   Costs from B to E

To E from	Through	Cost B to C	C to E	Total	Optimum
$B1$	$C1$	12	31	43	✕
$B1$	$C2$	15	32	47	
$B1$	$C3$	24	31	55	
$B1$	$C4$	28	30	58	
$B2$	$C1$	15	31	46	✕
$B2$	$C2$	16	32	48	
$B2$	$C3$	17	31	48	
$B2$	$C4$	21	30	51	
$B3$	$C1$	24	31	55	
$B3$	$C2$	17	32	49	
$B3$	$C3$	16	31	47	
$B3$	$C4$	15	30	45	✕
$B4$	$C1$	28	31	59	
$B4$	$C2$	21	32	53	
$B4$	$C3$	15	31	46	
$B4$	$C4$	12	30	42	✕

It is at this point that the benefits of dynamic programming begin to emerge. Hereafter, whenever the path involves point $C1$, we know immediately the optimum route and cost from $C1$ to the destination, and the nonoptimal paths are not considered.

Advancing (backwards) to the next stage examines costs when starting at the B points, as shown in Table 9-4. The first line, for example, shows the cost from B_1 to E as 43 which is composed of the cost from $B1$ to $C1$ of 12 plus the minimum cost from $C1$ to E of 31 which is taken from Table 9-3. Of the four possibilities of getting from $B1$ to E, the one passing through $C1$ is the minimum and the others are ruled out of further consideration.

The final calculation begins at the starting point A and evaluates the four possibilities as shown in Table 9-5.

The minimum cost from A to E, thus the minimum total cost, is 61. Tracing back through the tables, the optimum route is found to be A-$B3$-$C4$-$D3$-E.

Even though this is a simple problem, there seem to be a large number of calculations. Calling each line of Tables 9-2 through 9-5 a calculation, a total of 40 calculations were required. Compared to examining all of the possible routes between A and E, however,

Table 9-5   Costs from A **to** E

| To E from | Through | Cost | | | Optimum |
		A to B	B to E	Total	
A	$B1$	20	43	63	
A	$B2$	17	46	63	
A	$B3$	16	45	61	$\times$
A	$B4$	20	42	62	

we find that dynamic programming is efficient.   From A to B there are four possibilities, and from each of the B points there are four possibilities of passing to C, and similarly from C to D.   From D to E there is just one possibility.   The number of possible routes if all are considered, then, is $(4)(4)(4) = 64$.

The saving of effort would be more impressive if the problem had included another stage consisting of four positions.   The number of calculations by dynamic programming would have been the current 40 plus an additional 16 for a total of 56.   Examining all possible routes would require $(64)(4) = 256$ calculations.

Reviewing the key feature of dynamic programming: when once an optimal policy has been determined from an intermediate stage to the final stage, future calculations passing through that intermediate stage utilize only that optimal policy.

9-3   Constrained optimization   The method of dynamic programming will now be extended into the frequently encountered situation where some constraint exists.   In the process of this extension, we also apply dynamic programing to an actual thermal system—optimization of feedwater heating.

Heating the boiler feedwater with extraction steam, as shown in Fig. 9-2, improves the efficiency of a steam-power cycle and is a common practice in large central power stations.   Some plants utilize more than a half-dozen heaters which draw off extraction steam at as many different pressures.   That feedwater heaters improve the efficiency of a cycle can be shown by a calculation of a specific case, but a qualitative explanation may provide a better sense of this improvement.   First, we recall that in the steam-power cycle approximately 4 Btu of heat are supplied at the boiler for every Btu of work at the turbine shaft.   The difference of 3 Btu is the amount rejected at the condenser, which usually represents a loss. The proposal to try to use some of the heat rejected at the condenser

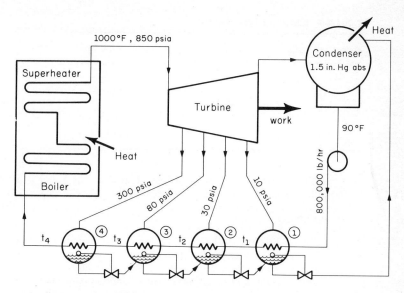

Fig. 9-2 Feedwater heater selection for Example 9-2.

for boiler heating is doomed, because, for example, if we tried to heat the feedwater with exhaust steam from the turbine, there would be no temperature difference between the exhaust steam and the feedwater to provide the driving force for heat transfer.

Extraction steam, however, has a higher temperature than exhaust steam, and can be used for the heating. Concentrating on a pound of extraction steam leaving the boiler, we find that it performs some work in the turbine before extraction then uses the remainder of its energy above saturated liquid at the condensing temperature to heat the feedwater. In effect, then, all of the heat supplied to that pound of steam in the boiler is eventually converted into work. The existence of feedwater heating by extraction steam raises the effectiveness of the cycle compared to rejecting 3 Btu of boiler heat per Btu of work.

It is further to be expected that the high-pressure steam is more valuable than the low-pressure steam because the steam extracted at high pressure would have had the capability of delivering additional work at the turbine shaft.

Example 9-2 An economic analysis has determined that 10,000 ft² of heat-transfer area should be utilized for feedwater heating in the steam-power cycle shown in Fig. 9-2. This 10,000 ft² can be distributed among four heaters which can be purchased in sizes from 1,000 to 10,000 ft² in increments of 1,000 ft². The overall heat-transfer coefficient of all heaters

Table 9-6   Extraction steam data

Extraction point and heater number	Saturation temperature, °F	Value of extraction steam, $/million Btu
1	193	0.1538
2	250	0.1887
3	312	0.2500
4	417	0.3119

is 500 Btu/(hr)(ft²)(°F). The cost of heat at the boiler is $0.40 per million Btu, and the values of extraction steam at the various extraction points as determined by thermodynamic calculations are listed in Table 9-6.

Solution   A qualitative prediction of the optimal distribution of area may at first suggest that all 10,000 ft² be placed at heater 1 where the lowest-cost extraction steam is available. Further reflection will show, however, that each additional increment of area at position 1 accomplishes progressively less heat exchange. If 1,000 ft² of area only raises the temperature of feedwater one degree at position 1, it would be preferable to use the area at some other position where it raises the temperature 10°F, for example, even though it uses more expensive steam. If the idea can be accepted that the area could conceivably be distributed through any or all of the positions, the brute-force method of optimization would require the calculation of many hundreds of different possibilities. Instead, the problem can be structured into a form adaptable to dynamic programming.

The first stage of calculation begins at heater 1 and explores all of its size possibilities—from zero to 10,000 ft². The outlet temperatures and heat transferred are calculated, using heat-exchanger performance equations in Sec. 4-13. The results are shown in Table 9-7.

Table 9-7   Calculations for first heater

Area, ft²	Outlet temperature °F	Saving, $/hr
0	90.0	0.00
1,000	137.87	9.43
2,000	163.49	14.47
3,000	177.20	17.17
4,000	184.54	18.62
5,000	188.47	19.39
6,000	190.58	19.80
7,000	191.70	20.03
8,000	192.31	20.15
9,000	192.63	20.21
10,000	192.80	20.24

As is typical of the first-stage calculations in dynamic programming, all of the possibilities are optimal since no possibilities can be ruled out yet. The move to the second feedwater heater introduces the unique feature which submits this constrained optimization problem to dynamic programming. Instead of thinking in terms of area in heater 2, the calculation considers *combined area* in the first two heaters. If 4,000 ft² is available in the first two heaters, for example, this stage of calculation determines the optimum distribution of that 4,000 ft². Having once found that optimal distribution, other combinations of 4,000 ft² in the first two heaters are ignored in future calculations.

If 10,000 ft² is available for the first two heaters, the optimal distribution can be selected from the 11 possible combinations shown in the first block of Table 9-8. Next, the optimal distribution of 9,000 ft² in the first two heaters is determined and results in the second block of Table 9-8. Similar blocks of calculations not shown in Table 9-8 are made for total area in the first two heaters from 8,000 ft² down to zero, and the optimum selected for each value of total area. The summary of this stage of calculation is presented in Table 9-9.

It is in Table 9-8 that dynamic programming first begins to pay off. Hereafter, when evaluating the effect of a certain value of total area in

Table 9-8 Optimal combinations of area in first two heaters

Total area in first two heaters, ft²	Area in heater No. 1	Outlet temperature, °F	Savings, $	Optimum
10,000	0	249.69	27.00	
10,000	1,000	249.60	28.31	
10,000	2,000	249.42	29.00	
10,000	3,000	249.08	29.32	
10,000	4,000	248.46	29.42	✕
10,000	5,000	247.30	29.34	
10,000	6,000	245.12	29.03	
10,000	7,000	241.06	28.37	
10,000	8,000	233.47	27.11	
10,000	9,000	219.29	24.72	
10,000	10,000	192.80	20.24	
9,000	0	249.42	26.95	
9,000	1,000	249.24	28.26	
9,000	2,000	248.91	28.91	
9,000	3,000	248.29	29.19	
9,000	4,000	247.12	29.20	✕
9,000	5,000	244.95	28.94	
9,000	6,000	240.89	28.31	
9,000	7,000	233.30	27.06	
9,000	8,000	219.12	24.68	
9,000	9,000	192.63	20.21	

Table 9-9 Optimal distribution of area in first two heaters

Total area in first two heaters, ft^2	Area in heater No. 1	Outlet temperature, °F	Savings, $
0	0	90.00	0.00
1,000	0	164.36	12.57
2,000	0	204.16	19.30
3,000	1,000	217.87	22.95
4,000	1,000	232.80	25.48
5,000	2,000	236.73	26.85
6,000	2,000	242.90	27.90
7,000	3,000	244.02	28.47
8,000	3,000	246.80	28.94
9,000	4,000	247.12	29.20
10,000	4,000	248.46	29.42

the first two heaters, only the optimal distribution of area between the first two heaters is considered, and nonoptimal distributions are ignored.

The next stage of calculation examines each possibility of total area from zero to 10,000 in the *first three* heaters. The optimal distribution is found by examining all the possibilities of area placement in heater 3. The difference between the total area and the area in heater 3 is that available for the first two heaters, for which Table 9-9 indicates the best distribution. The summary of this stage of calculation is presented in Table 9-10.

The final calculation stage determines the area in heater 4. In this calculation only one total area is considered, namely, 10,000 ft^2. Table 9-11 shows the result of this calculation.

Table 9-10 Optimal distribution of area in first three heaters

Total area in first three heaters, ft^2	Area in first two heaters, ft^2	Outlet temperature, °F	Savings, $
0	0	90.00	0.00
1,000	1,000	164.36	12.57
2,000	1,000	232.97	20.80
3,000	2,000	254.28	25.31
4,000	2,000	281.10	28.53
5,000	3,000	285.03	31.01
6,000	3,000	297.57	32.51
7,000	4,000	299.85	33.52
8,000	5,000	300.46	34.50
9,000	5,000	305.82	35.14
10,000	6,000	306.33	35.51

Table 9-11 Optimal area in fourth heater

Area in heater 4, ft^2	Area in first three heaters, ft^2	Outlet temperature, °F	Total savings, $	
10,000	0	416.37	23.05	
9,000	1,000	416.09	30.31	
8,000	2,000	415.76	33.68	
7,000	3,000	414.95	36.63	
6,000	4,000	413.80	37.88	
5,000	5,000	411.20	39.90	
4,000	6,000	407.20	40.24	
3,000	7,000	399.04	40.51	→ optimum
2,000	8,000	383.61	40.36	
1,000	9,000	357.49	38.78	
0	10,000	306.33	35.51	

The optimal distribution, as shown by Table 9-11, is to place 3,000 ft^2 in heater 4, leaving 7,000 ft^2 for the first three heaters. Tracing back through Tables 9-10, 9-9, and 9-7 defines the area in the first three heaters:

Area in heater 4 = 3,000 ft^2
Area in heater 3 = 3,000 ft^2
Area in heater 2 = 3,000 ft^2
Area in heater 1 = 1,000 ft^2

with the optimal saving of $40.51 and an outlet feedwater temperature from the chain of 399.04°F.

Example 9-2 began with the stipulation that 10,000 ft^2 of heat-transfer area was available for feedwater heating, and the question may arise how this figure could be specified before an optimal distribution was determined. The question is valid and suggests that Example 9-2 is a suboptimization within the larger optimization of determining the total area to be made available for feedwater heating.

Another specification in Example 9-2 was that the heat-exchange area was available in 1,000-ft^2 increments. This is a coarser increment than actually available, so a recalculation could offer area in 100-ft^2 increments, for example, examining possible areas 500 ft^2 on either side of the 1,000-3,000-3,000-3,000-ft^2 distribution found with the 1,000-ft^2 increments.

9-4 Further applications of dynamic programming Dynamic programming is based on a simple concept, although the execution in complex situations may be tricky from the standpoint of logic.

Dynamic programming finds application especially in certain staged processes in engineering, where the decisions made at one stage influence the input conditions at the next stage.

BIBLIOGRAPHY

Bellman, R. E.: "Dynamic Programming," Princeton University Press Princeton, N.J., 1957.
Bellman, R., and S. Dreyfus: "Applied Dynamic Programming," Princeton University Press, Princeton, N.J., 1962.
Nemhauser, G. L.: "Introduction to Dynamic Programming," John Wiley & Sons, Inc., New York, 1960.
Roberts, S.: "Dynamic Programming in Chemical Engineering and Process Control," Academic Press, Inc., 1964.

PROBLEMS

9-1. Using dynamic programming, determine the flight plan for a commercial airliner flying between two cities 600 miles apart such that the minimum fuel is consumed during the flight. Specifically, the altitudes at locations B through F in Fig. 9-3 during the course of the flight are to be specified.

Table 9-12 shows the fuel consumption for 100-mile ground distances as a function of the climb or descent during that distance. Determine the flight plan and the minimum fuel cost.

Ans.: 930 gal.

9-2. A minimum-cost pipeline is to be constructed between positions A and G in Fig. 9-4 and can pass through any of six locations in the successive stages B, C, D, E, and F.

(a) If all possible combinations of routes are investigated, how many different paths must be examined?

(b) If dynamic programming is used, how many calculations must be

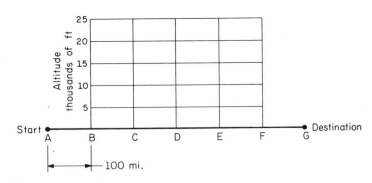

Fig. 9-3 Altitudes and distances in Prob. 9-1.

Table 9-12   Fuel consumption per 100 miles of ground travel, gal

From: Altitude, ft ╲ To: Altitude, ft	0	5,000	10,000	15,000	20,000	25,000
0	...	500	600	690	770	840
5,000	100	200	335	500	590	760
10,000	40	60	100	280	387	575
15,000	0	20	40	67	195	380
20,000	0	0	10	30	60	200
25,000	0	0	0	0	20	30

made where one calculation consists of evaluating the cost from the stage in question to the terminal point?

Ans.: 7,776, 156.

9-3. The maintenance and improvement schedule for a plant is to be planned so that a maximum total profit will be achieved during the 5-year life of the plant. Since there is a carryover of the condition of the plant from one year to the next, a qualitative expectation of the best maintenance plan is that of heavy expenditures early in the life and permitting the plant to decline in the last few years, since there is no salvage value of this particular plant.

The current income level of the plant is $15,000 per year and Table 9-13 shows how various expenditures for maintenance and improvements alter the income level for the next year. For example, with the current income level of $15,000 an expenditure of $15,000 for maintenance and improvements increases the income level to $20,000 leaving a profit for the first year of $20,000 − $15,000 = $5,000 and an income level at the start of the second year of $20,000.

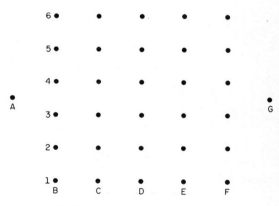

Fig. 9-4   Pipeline route in Prob. 9-2.

Table 9-13   **New annual income levels (in thousands of dollars)**

Annual expenditures for maintenance and improvements	*Current annual income level (in thousands of dollars)*								
	0	5	10	15	20	25	30	35	40
0	0	0	5	10	15	20	20	25	30
5	5	10	10	15	20	25	30	30	35
15	10	15	15	20	25	30	35	35	40
30	15	20	20	25	30	35	40	40	40

Table 9-13 roughly reflects the fact that when the income is low, indicating a poor condition of the plant, expenditures are effective in bringing up the income level because what repairs are needed are obvious. When the income level is already high the maintenance expenditures are heavily preventive—some of them unnecessary.

Using dynamic programming, determine the 5-year maintenance plan that results in maximum total profit for the 5 years.

Ans.: $30,000, $15,000, $5,000, $5,000, and $5,000—for a total profit of $85,000.

9-4. An underground electrical conductor, shown in Fig. 9-5, is 6,000 ft long and is to be cooled by enclosing it in a conduit through which chilled oil flows. The conduit in turn is surrounded by insulation to reduce the heat gain from the ground, which is at a temperature of 50°F.

The insulation is available in thicknesses of 1, 2, and 3 in. The oil rises in temperature due to heat from both the conductor and from the ground.

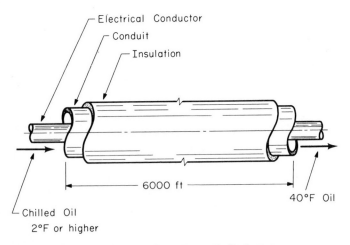

Fig. 9-5   Cooled underground conductor in Prob. 9-4.

Table 9-14 Inlet-oil temperatures resulting in given outlet temperatures, °F

Outlet temperature from a 1,000-ft section, °F

Insulation thickness, in.	40	39	38	37	36	35	34	33	32	31	30	29	28	27	26	25	24	23	22
1	35	34	33	31	30	29	27	26	25	24	22	21	20	18	17	16	15	13	12
2	36	35	34	33	31	30	30	29	27	26	25	23	23	22	21	19	18	17	16
3	37	36	35	34	33	32	30	29	28	27	26	25	24	23	22	21	20	18	17
	21	20	19	18	17	16	15	14	13	12	11	10	9	8	7	6			
1	11	9	8	7	6	4	3	2	0	−1	−2	−3	−5	−6	−7	−9			
2	15	14	13	11	9	9	8	7	6	5	3	2	1	0	−1	−2			
3	16	15	14	13	12	11	9	9	7	6	5	4	3	2	1	0			

The heat from the ground is a function of the temperature of the oil in the section and the amount of insulation.   Table 9-14 shows inlet-oil temperatures at given outlet temperatures and thicknesses of insulation for 1,000-ft sections.

Using dynamic programming with the 6,000-ft length divided into 1,000-ft sections, determine the thickness of insulation in each section that results in the minimum total consumption of insulation with a 40°F outlet-oil temperature and an inlet temperature of 2°F or higher.

Ans.: 3, 2, 2, 1, 1, 1 in.

10

Geometric Programming

10-1 Introduction Geometric programming is one of the newest methods of optimization. Clarence Zener first recognized the significance of the geometric and arithmetic mean in cases of unconstrained optimization. Since then, Duffin, Peterson, Wilde and Passy have all extended the methods to accommodate more general optimization problems. The form of problem statement that is particularly adaptable to treatment by geometric programming is a sum of polynomials for both the objective function and the constraint equations. These polynomials can be made up of combinations of variables to either positive or negative noninteger exponents. After seeing the utility of polynomial representations in Chap. 4, the ability to optimize such functions is clearly recognized to be of engineering importance.

A feature of geometric programming is that the first stage of the solution is to find the optimum value of the function, rather than to first determine the values of the independent variables that give this optimum. This knowledge of the optimum value may be all that is of interest and the calculation of the values of the variables can be omitted.

The ideas behind geometric programming may at first strike the reader as a dark art. Use of the method soon dispels this feeling. Rather than starting with the completely general solution, we shall begin with simple cases. The factors that influence the simplicity or complexity are the number of variables, the number of terms in the objective function, whether or not constraints exist, whether the constraints are equality or inequality constraints and whether they are mixed in sense, and finally, whether the terms in the objective function are all positive.

10-2 Types of problems that can be solved by geometric programming

An example of the statement of an optimization problem that can be attacked by geometric programming is to minimize

$$y = x_1 x_2 + 3x_3 + \frac{4}{x_1 x_3} + \frac{5}{\sqrt{x_2}}$$

subject to the constraints

$$x_1 + \frac{x_2}{x_3} = 5$$

$$x_2{}^2 + 3x_3 = 1$$

Since the x variables are normally understood to represent physical quantities, a further constraint that is understood may be that x_1, x_2, and x_3 are all equal to or greater than zero.

In a more general form, the optimization problem with equality constraints can be stated as optimizing

$$y = \sum_{t=1}^{T} c_t \prod_{n=1}^{N} x_n{}^{a_{tn}}$$

under the constraints

$$\sum_{t=1}^{T_j} c_{jt} \prod_{n=1}^{N} x_{jn}{}^{a_{jtn}} = f_j \qquad j = 1 \cdots M$$

where T = number of terms in the objective function and constraints (eight in the case of the above example)

N = number of variables (three in the example)

c_t = coefficient of the t term

a_{tn} = exponent of x_n in term t

T_j = number of terms in constraint equation j
M = number of constraint equations
c_{jt} = coefficient of t term in j constraint
a_{jtn} = exponent of x_n in t term of j constraint

and $\prod$ indicates multiplication

10-3 Degree of difficulty

Duffin, Peterson, and Zener[1] define "degree of difficulty" when applied to geometric programming problems as $\sum T - (N + 1)$. The lowest degree of difficulty is zero, which means that the sum of terms in the objective function and constraint equations is 1 greater than the number of independent variables. The classes of problems that this chapter will explain are, in order:

1. Unconstrained optimization, one independent variable, zero degree of difficulty
2. Unconstrained optimization, several independent variables, zero degree of difficulty
3. Unconstrained optimization, one independent variable, degree of difficulty = 1
4. Constrained optimization, several independent variables, zero degree of difficulty

10-4 One variable with zero degree of difficulty

A specific illustration will demonstrate the procedure of geometric programming.

Example 10-1 Determine the optimum pipe diameter which results in minimum first-plus-operating cost for 100 ft of pipe conveying a given water-flow rate. The first cost in dollars of the pipe, installed, is $150D$ where D is the pipe diameter in inches. The lifetime pumping cost is $122{,}500/D^5$ dollars. It is to be expected that the pumping cost will be proportional to D^{-5}, because the pumping cost for a specified flow rate, number of hours of operation, pump efficiency, motor efficiency, and electric rate is proportional to the pressure drop in the pipe. Further, this pressure drop Δp is

$$\Delta p = f \frac{L}{D} \frac{V^2}{2g_c} \rho = f \frac{L}{D} \left(\frac{Q}{\pi D^2/4} \right)^2 \frac{\rho}{2g_c} = \frac{\text{constant}}{D^5}$$

The objective function, the cost y, in terms of the variable x is

$$y = 150x + \frac{122{,}500}{x^5} \tag{10-1}$$

[1] R. G. Duffin, E. L. Peterson, and C. M. Zener, "Geometric Programming," John Wiley & Sons, Inc., New York, 1967.

Solution   To provide a check on the geometric-programming method, optimize by calculus.

$$\frac{dy}{dx} = 150 - \frac{(5)(122{,}500)}{x^6} = 0$$

$$x^* = 4 \text{ in.} \quad \text{and} \quad y^* = \$720$$

This problem has zero degree of difficulty, since $T = 2$ and $N = 1$, so $T - (N + 1) = 0$.

To lay the groundwork for the optimization by geometric programming, rewrite Eq. (10-1) as

$$y = u_1 + u_2 = c_1 x^{a_1} + c_2 x^{a_2} \tag{10-2}$$

where, in this case,

$$u_1 = 150x \qquad u_2 = \frac{122{,}500}{x^5} \qquad c_1 = 150$$

$$c_2 = 122{,}500 \qquad a_1 = 1 \quad \text{and} \quad a_2 = -5$$

Next, fabricate a function g such that

$$g = \left(\frac{u_1}{w_1}\right)^{w_1} \left(\frac{u_2}{w_2}\right)^{w_2} = \left(\frac{c_1 x^{a_1}}{w_1}\right)^{w_1} \left(\frac{c_2 x^{a_2}}{w_2}\right)^{w_2} \tag{10-3}$$

where

$$w_1 + w_2 = 1 \tag{10-4}$$

A certain combination of values of w_1 and w_2 will provide a maximum value of g. To determine these values of w_1 and w_2, apply the method of Lagrange multipliers to Eq. (10-3) subject to the constraint of Eq. (10-4). The maximum values of g and $\ln g$ both occur at the same value of the w's, and it is more convenient to optimize $\ln g$. Maximize

$$\ln g = w_1(\ln u_1 - \ln w_1) + w_2(\ln u_2 - \ln w_2) \tag{10-5}$$

subject to

$$w_1 + w_2 - 1 = \phi = 0 \tag{10-6}$$

Using the method of Lagrange multipliers,

$$\nabla(\ln g) - \lambda \nabla \phi = 0$$

$$\phi = 0$$

which provides the three equations

$$w_1: \quad \ln u_1 - 1 - \ln w_1 - \lambda = 0$$

$$w_2: \quad \ln u_2 - 1 - \ln w_2 - \lambda = 0$$

$$w_1 + w_2 - 1 = 0$$

The unknowns are w_1, w_2, and λ, and the solutions for w_1 and w_2 are

$$w_1 = \frac{u_1}{u_1 + u_2} \tag{10-7}$$

and

$$w_2 = \frac{u_2}{u_1 + u_2} \tag{10-8}$$

Substituting these values of w_1 and w_2 into Eq. (10-3) gives

$$g = \left[\frac{u_1}{u_1/(u_1 + u_2)} \right]^{u_1/(u_1+u_2)} \left[\frac{u_2}{u_2/(u_1 + u_2)} \right]^{u_2/(u_1+u_2)}$$

Thus,

$$g = u_1 + u_2 \tag{10-9}$$

Pausing at this point to assess our status, by the choice of w_1 and w_2 according to Eqs. (10-7) and (10-8) the value of g is made equal to that of $u_1 + u_2$ and, therefore, also to y. Any other combination of w_1 and w_2 results in a value of $g < (u_1 + u_2) = y$. We remember that our original objective is to minimize y, so the next step is to use the value of x in Eq. (10-9) that results in the minimum value of $u_1 + u_2$. The value of g at this condition g^* will, therefore, be the one where the w's are chosen such that g always equals y, but also with the value of x chosen so that y is a minimum. This value of x^* can be found by equating the derivative of Eq. (10-2) to zero.

$$a_1 c_1 x^{(a_1-1)} + a_2 c_2 x^{(a_2-1)} = 0$$

Multiplying by x,

$$a_1 c_1 x^{a_1} + a_2 c_2 x^{a_2} = 0$$

so

$$a_1 u_1^* + a_2 u_2^* = 0 \tag{10-10}$$

where u_1^* and u_2^* are the values of u_1 and u_2 at the minimum value of y.

From Eq. (10-10),

$$u_1^* = - \frac{a_2 u_2^*}{a_1}$$

which, when substituted into Eqs. (10-7) and (10-8), yields

$$w_1 = \frac{-(a_2/a_1) u_2^*}{-(a_2/a_1) u_2^* + u_2^*} = \frac{-a_2}{a_1 - a_2} \tag{10-11}$$

and

$$w_2 = \frac{u_2^*}{-(a_2/a_1)u_2^* + u_2^*} = \frac{a_1}{a_1 - a_2} \tag{10-12}$$

When these values of w_1 and w_2 are substituted into Eq. (10-3), the value of g^* is found to be

$$g^* = \left[\frac{c_1 x^{a_1}}{-a_2/(a_1 - a_2)} \right]^{-a_2/(a_1-a_2)} \left[\frac{c_2 x^{a_2}}{a_1/(a_1 - a_2)} \right]^{a_1/(a_1-a_2)}$$

Of particular significance is the fact that x has canceled out of the expression for y^*.

$$g^* = \left(\frac{c_1}{w_1}\right)^{w_1} \left(\frac{c_2}{w_2}\right)^{w_2} = y^* \tag{10-13}$$

In the execution of geometric programming, this property will be used as a convenient way of determining the values of w. They will be proportioned in the expression for g such that the x's cancel out.

Example 10-2   Solve Example 10-1 by geometric programming.

$$y = 150x + \frac{122,500}{x^5}$$

Solution

$$y^* = 150x^* + \frac{122,500}{x^{*5}} = g^* = \left(\frac{150x}{w_1}\right)^{w_1} \left(\frac{122,500}{x^5 w_2}\right)^{w_2}$$

provided that w_1 and w_2 are chosen such that

$$a_1 w_1 + a_2 w_2 = w_1 - 5w_2 = 0$$

and

$$w_1 + w_2 = 1$$

Solving,

$$w_1 = \frac{5}{6} \quad \text{and} \quad w_2 = \frac{1}{6}$$

Substituting these values for w_1 and w_2 into the expression for g^* results in the cancellation of the x's, leaving

$$y^* = g^* = \left(\frac{150}{5/6}\right)^{5/6} \left(\frac{122,500}{1/6}\right)^{1/6} = (180)^{5/6}(735,000)^{1/6} = 720$$

The value of x^* can be found by applying Eq. (10-7)

$$w_1 = \frac{5}{6} = \frac{u_1^*}{u_1^* + u_2^*} = \frac{u_1^*}{y^*} = \frac{150x}{720}$$

So $x^* = 4 =$ optimum diameter of 4 in. and $y^* = \$720 =$ minimum cost.

10-5 Arithmetic and geometric mean   Several observations are now appropriate: first a mathematical one, then a physical.   The maximum value of g equals y, so nonmaximum values of g are less than y. Thus,

$$y \geq g$$

and

$$u_1 + u_2 \geq \left(\frac{u_1}{w_1}\right)^{w_1}\left(\frac{u_2}{w_2}\right)^{w_2}$$

Defining $U_1 = u_1/w_1$ and $U_2 = u_2/w_2$,

$$w_1 U_1 + w_2 U_2 \geq (U_1)^{w_1}(U_2)^{w_2} \tag{10-14}$$

where

$$w_1 + w_2 = 1$$

Equation (10-14) is the Cauchy inequality.   It states that the arithmetic mean is equal to or greater than the geometric mean. The use of the geometric mean in this method of optimization explains the origin of the name *geometric programming*.

Another observation from geometric programming is that the optimum occurs when the contributing terms are in a certain relation to the total.   Using Eqs. (10-7) and (10-11) at the optimum,

$$w_1 = \frac{u_1^*}{u_1^* + u_2^*} = \frac{-a_2}{a_1 - a_2}$$

The term w_1 is the weighting function that expresses what fraction u_1^* is of the total y^*.   Furthermore, this fraction is dependent upon the magnitudes of the exponents a_1 and a_2.

Example 10-2 illustrates some physical insight, which adds some utility to the geometric program beyond obtaining the optimum conditions.   The two terms u_1 and u_2 in the objective function are the two contributors to the total cost—that of the pipe and that of the operating cost, respectively.   At the minimum total cost, the cost of the pipe is $5/6$ of the total cost and the operating expense is $1/6$ of the total.   These fractions are dictated by the exponents of x which are unity and -5.

What would be the new optimum condition if the cost of the pipe goes up from $150D$ to $200D$?   The contribution of the pipe to the total cost remains $5/6$ of the total.   The total cost at the optimum will now be higher, the cost of the pipe will be more, and the optimum pipe size will be less, but the fraction of the total cost contributed by the pipe remains constant.

The above observation has useful qualitative value in economic analyses where costs are constantly changing. The distribution of costs among various contributors remains constant even if the prices change, so long as the functional expression (the exponents of x) remains fixed.

10-6 Unconstrained, multivariable optimization of zero degree of difficulty The demonstration of optimization by geometric programming in Sec. 10-4 was performed on a function of a single variable. We next generalize to the multivariable case where $N > 1$, although retaining $T - (N + 1)$ equals zero. We shall find that the same procedures apply in the multivariable case. In the execution of geometric programming, the weighting functions are chosen so that all of the x variables cancel out in the expression for g.

Given a function y to be minimized,

$$y = \sum_{t=1}^{T} c_t \prod_{n=1}^{N} x_n{}^{a_{tn}} = \sum_{t=1}^{T} u_t \tag{10-15}$$

we fabricate a function g such that

$$g = \prod_{t=1}^{T} \left(\frac{u_t}{w_t}\right)^{w_t} \tag{10-16}$$

where

$$\sum_{t=1}^{T} w_t = 1 \tag{10-17}$$

To determine the combination of w_t's that yield the maximum value of g, apply the method of Lagrange multipliers on $\ln g$, subject to the constraint of Eq. (10-17)

$$\nabla(\ln g) - \lambda \nabla \sum_{t=1}^{T} w_t = 0 \tag{10-18}$$

where the gradient vector in Eq. (10-18) represents T scalar equations

$$\ln u_t - 1 - \ln w_t - \lambda = 0 \qquad t = 1, \ldots, T \tag{10-19}$$

Taking the antilog of Eqs. (10-19) gives

$$w_1 = u_1 e^{-(\lambda+1)}$$
$$w_2 = u_2 e^{-(\lambda+1)}$$
$$\cdot \cdot \cdot \cdot \cdot \cdot \cdot \cdot$$
$$w_T = u_T e^{-(\lambda+1)}$$

Summing these equations and equating to the constraint equation, Eq. (10-17), gives

$$\sum_{t=1}^{T} w_t = e^{-(\lambda+1)} \sum_{t=1}^{T} u_t = e^{-(\lambda+1)}y = 1$$

Then

$$e^{-(\lambda+1)} = \frac{1}{\sum u_t}$$

and

$$w_t = \frac{u_t}{\sum u_t} \qquad \text{for maximum } g \qquad\qquad (10\text{-}20)$$

Substituting these values of w_t back into Eq. (10-16) shows that

$$g = \prod_{t=1}^{T} \left(\frac{u_t}{u_t / \sum u_t} \right)^{u_t / \Sigma u_t} = \sum u_t = y$$

Having succeeded in choosing w's such that $g = y$, next choose the x's that give the minimum y and, thus, g. From Chap. 7 for unconstrained optimization, we recall that the optimum occurs where

$$\nabla y = 0 \qquad\qquad (10\text{-}21)$$

or

$$\frac{\partial y}{\partial x_1} = 0 \qquad \frac{\partial y}{\partial x_2} = 0 \qquad \cdots \qquad \frac{\partial y}{\partial x_N} = 0$$

Using the polynomial representation in Eq. (10-15), Eq. (10-21) becomes

$$\sum_{t=1}^{T} c_t a_{tn} x_n^{a_{tn}-1} \prod_{j=1}^{n-1} x_j^{a_{tj}} \prod_{n+1}^{N} x_j^{a_{tj}} = 0 \qquad n = 1, \ldots, N$$

Multiplying by x_n,

$$\sum_{t=1}^{T} c_t a_{tn} \prod_{n=1}^{N} x_n^{a_{tn}} = 0$$

or

$$\sum_{t=1}^{T} a_{tn} u_t^* = 0 \qquad n = 1, \ldots, N \qquad\qquad (10\text{-}22)$$

From Eq. (10-20),

$$u_t = w_t \sum u_t \qquad t = 1, \ldots, T$$

which when substituted into the optimum condition of Eq. (10-22) results in

$$\sum_{t=1}^{T} a_{tn} w_t \sum u_t^* = 0$$

so at the optimum

$$\sum_{t=1}^{T} a_{tn} w_t = 0 \qquad n = 1, \ldots, N \tag{10-23}$$

which is the requirement that the w's be chosen such that the x's cancel out in Eq. (10-16). Equation (10-23) represents N individual equations which in combination with Eq. (10-17) give $N + 1 = T$ equations, permitting the solution of the T unknown values of w_t.

Substituting the values of w_t according to Eq. (10-23) into Eq. (10-16) provides $g^* = y^*$, the minimum value of y.

Example 10-3   The pump and piping of Example 10-1 are actually part of a waste-treatment system, as shown in Fig. 10-1. The system accomplishes the treatment by a combination of dilution and chemical action in order that the effluent meet code requirements. The greater the amount of dilution, the smaller the required size of the reactor. Its cost is $432/Q$ dollars where Q is the rate of water flow in cfs. The equation for the pumping cost with Q broken out of the combined constant is $(972,000)Q^2/D^5$.

Optimize the total system.

Solution   The total cost is the sum of the costs of the pipe, the pumping power, and the treatment plant.

$$y = 150D + \frac{972,000Q^2}{D^5} + \frac{432}{Q}$$

$$y^* = g^* = \left(\frac{150}{w_1}\right)^{w_1} \left(\frac{972,000}{w_2}\right)^{w_2} \left(\frac{432}{w_3}\right)^{w_3}$$

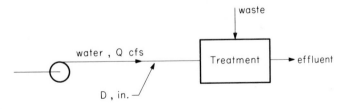

Fig. 10-1   Waste-treatment system in Example 10-3.

provided that

$$\sum_{t=1}^{3} w_t = 1$$

and

$$\sum_{t=1}^{3} a_{tn}w_t = 0 \qquad n = 1, 2$$

Written out, these equations are

$$w_1 + w_2 + w_3 = 1$$
$$D: \qquad w_1 - 5w_2 = 0$$
$$Q: \qquad 2w_2 - w_3 = 0$$

Solving,

$$w_1 = \frac{5}{8} \qquad w_2 = \frac{1}{8} \qquad w_3 = \frac{2}{8}$$

Then

$$y^* = \left(\frac{150}{5/8}\right)^{5/8} \left(\frac{972,000}{1/8}\right)^{1/8} \left(\frac{432}{2/8}\right)^{2/8} = \$1,440$$

$u_1^* = 150D^* = w_1(1,440) = 900,$ so $D^* = 6$ in. $u_3^* = 432/Q^* = w_3(1,440) = 360,$ so $Q^* = 1.2$ cfs.

The core of the execution of Example 10-3 by geometric programming was the solution of three simultaneous *linear* equations for the w's. Had the problem been solved by calculus, two simultaneous *nonlinear* equations (from $\nabla y = 0$) would have been solved. In general, when the degree of difficulty is zero, solution by geometric programming requires solution of one more equation in the set of simultaneous equations than required by calculus, *but the equations are linear.*

An evaluation of the utility of geometric programming at this stage is, then, when the degree of difficulty is zero the optimization by geometric programming is likely to be much simpler than by calculus. A further benefit of geometric programming is that it immediately shows the optimum distribution between the various terms of the objective function.

10-7 Unconstrained optimization, degree of difficulty greater than zero In the objective functions considered thus far, T was equal to $N + 1$, so the situations examined had zero degrees of difficulty. We next consider the case where the degree of difficulty is greater than zero.

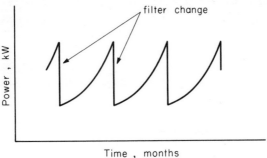

Fig. 10-2 Power consumption in Example 10-4.

Example 10-4 As the filters in an air-handling system become dirty, the fan
power increases, as shown in Fig. 10-2 and expressed by the equation

$$\text{Power } (kW) = 150 + 30t + 9t^2$$

where t = time since last filter change, months
The cost of electricity is \$0.012 per kWh and the system operates 500
hr/month. The cost of a filter change is \$450.
(a) What is the minimum annual total cost of power plus filters?
(b) What is the optimum time in months between filter changes?

Solution A qualitative observation is that some optimum time between filter
changes is expected, because a long period between changes results in
high power costs, but frequent changes cause an unnecessarily high cost
of filters.

Annual costs:
Filters:

$$(\$450 \text{ per change}) \left[\frac{12}{T} \text{ (changes/year)} \right] = \frac{5{,}400}{T} \text{ (\$/year)}$$

where T = time between changes, months
Power: the operating cost per period is

$$(\$0.012 \text{ per kWh})(500 \text{ hr/month}) \int_0^T (150 + 30t + 9t^2)\, dt$$

$$= 6(150T + 15T^2 + 3T^3)$$

so the annual power cost is the cost per period multiplied by the number
of periods per year.

$$\left[\frac{12}{T} \text{ (periods/year)} \right] (900T + 90T^2 + 18T^3)$$

Total:

$$10{,}800 + \frac{5{,}400}{T} + 1{,}080T + 216T^2$$

The constant does not affect the optimum value of T, so the objective function becomes

$$y = \frac{5,400}{T} + 1,080T + 216T^2$$

In this problem, there are three terms and one independent variable, so the degree of difficulty is 1. Proceeding as far as possible with the principles of geometric programming previously explained,

$$y = g = \left(\frac{5,400}{w_1}\right)^{w_1} \left(\frac{1,080}{w_2}\right)^{w_2} \left(\frac{216}{w_3}\right)^{w_3} \tag{10-24}$$

where

$$w_1 + w_2 + w_3 = 1 \tag{10-25}$$

and in order for the T's to cancel

$$-w_1 + w_2 + 2w_3 = 0 \tag{10-26}$$

It is here that the first difficulty appears, because the two equations, Eqs. (10-25) and (10-26), are inadequate to solve for all three values of w. It is possible however, to solve for two of the w's in terms of the remaining w. For example,

$$w_1 = \frac{w_3 + 1}{2} \quad \text{and} \quad w_2 = \frac{-3w_3 + 1}{2} \tag{10-27}$$

Substituting these values into Eq. (10-24) yields

$$y = g = \left[\frac{5,400}{(w_3 + 1)/2}\right]^{(w_3+1)/2} \left[\frac{1,080}{(-3w_3 + 1)/2}\right]^{(-3w_3+1)/2} \left(\frac{216}{w_3}\right)^{w_3} \tag{10-28}$$

It now becomes evident that Eq. (10-24) represents only a partial optimization and that the final stage of optimization requires determination of the value of w_3 that gives the minimum value of g. Equation (10-28) could be differentiated with respect to w_3 and equated to zero, but it is more convenient to find the value of w_3 that gives the minimum value of $\ln g$.

$$\ln g = -\frac{w_3 + 1}{2} \ln\left[\frac{(w_3 + 1)/2}{5,400}\right] - \frac{(-3w_3 + 1)}{2} \ln\left[\frac{(-3w_3 + 1)/2}{1,080}\right]$$
$$- w_3 \ln\frac{w_3}{216}$$

Differentiating,

$$\frac{d (\ln g)}{d w_3} = -\frac{1}{2} \ln\left[\frac{(w_3 + 1)/2}{5,400}\right] - 5,400\left(\frac{\frac{1}{2}}{5,400}\right)$$
$$+ \frac{3}{2} \ln\left[\frac{(-3w_3 + 1)/2}{1,080}\right] + 1,080\left(\frac{\frac{3}{2}}{1,080}\right)$$
$$- \ln\frac{w_3}{216} - 216\left(\frac{1}{216}\right) \tag{10-29}$$

The constant terms in Eq. (10-29) cancel and they will cancel in all geometric-programming problems of unity degree of difficulty. The remaining terms can be written

$$\ln \left\{ \left[\frac{5,400}{(w_3 + 1)/2} \right]^{\frac{1}{2}} \left[\frac{1,080}{(-3w_3 + 1)/2} \right]^{-\frac{3}{2}} \left(\frac{216}{w_3} \right) \right\} = 0$$

or

$$\left[\frac{5,400}{(w_3 + 1)/2} \right]^{\frac{1}{2}} \left[\frac{1,080}{(-3w_3 + 1)/2} \right]^{-\frac{3}{2}} \left(\frac{216}{w_3} \right) = 1 \qquad (10\text{-}30)$$

Equation (10-30) is a nonlinear equation which can be solved for w_3 by the Newton-Raphson method. The result of this solution is $w_3 = 0.113$, which can be substituted into Eq. (10-27) to find $w_1 = 0.557$ and $w_2 = 0.330$. The minimum values of y and g, then, are from Eq. (10-24)

$$y^* = g^* = \left(\frac{5,400}{0.557} \right)^{0.557} \left(\frac{1,080}{0.330} \right)^{0.330} \left(\frac{216}{0.113} \right)^{0.113} = \$5,620$$

To find the optimum value of T

$$u_2^* = w_2(5,620) = 1,080T$$

so $T^* = 1.72$ months, and annual cost $= 10,800 + 5,620 = \$16,420$.

After having gone once through a solution such as Example 10-4, it is possible to jump directly from Eq. (10-27) to Eq. (10-30). The left side of Eq. (10-30) is identical to the right side of Eq. (10-24), except that the exponents are the coefficients of the w_3 terms in the equations for w_1 and w_2. The exponent of the w_3 term is unity.

It is appropriate to make another evaluation of the utility of geometric programming. In Example 10-4 where the degree of difficulty was 1, one nonlinear equation, which was a rather messy equation, had to be solved. If the degree of difficulty is 2, then two simultaneous nonlinear equations must be solved for the optimal w's. Geometric programming may result in a considerable saving of effort in comparison to the calculus method when the degree of difficulty is zero, but for a degree of difficulty greater than zero, it may demand more effort than optimization by calculus.

10-8   Constrained optimization with zero degree of difficulty   The final type of geometric-programming problem to be explored is one with an equality constraint. Only the zero degree of difficulty case will be considered, and the total number of terms T means the sum of those in the objective function and the constraint. Suppose that the objective function to be minimized is

$$y = u_1 + u_2 + u_3 \qquad (10\text{-}31)$$

subject to the constraint

$$u_4 + u_5 = 1 \qquad (10\text{-}32)$$

where the u's are polynomials in terms of four independent variables x_1, x_2, x_3, and x_4. It is necessary that the right side of the constraint equation be unity, which poses no problem as long as one pure numerical term appears in the equation. If that number is not unity, the entire equation can be divided by the number to convert it to unity.

The objective function can be rewritten

$$y = g = \left(\frac{u_1}{w_1}\right)^{w_1} \left(\frac{u_2}{w_2}\right)^{w_2} \left(\frac{u_3}{w_3}\right)^{w_3} \tag{10-33}$$

provided that

$$w_1 + w_2 + w_3 = 1 \tag{10-34}$$

and

$$w_i = \frac{u_i}{u_1 + u_2 + u_3} \tag{10-35}$$

The constraint equation can also be rewritten as

$$u_4 + u_5 = 1 = \left(\frac{u_4}{w_4}\right)^{w_4} \left(\frac{u_5}{w_5}\right)^{w_5} \tag{10-36}$$

provided that

$$w_4 + w_5 = 1 \tag{10-37}$$

and

$$w_4 = \frac{u_4}{1} = u_4 \quad \text{and} \quad w_5 = \frac{u_5}{1} = u_5 \tag{10-38}$$

Equation (10-36) can be raised to the M power, where M is an arbitrary constant, and its value remains unity.

$$1 = \left(\frac{u_4}{w_4}\right)^{Mw_4} \left(\frac{u_5}{w_5}\right)^{Mw_5} \tag{10-39}$$

Next multiply Eq. (10-33) by Eq. (10-39)

$$y = g = \left(\frac{u_1}{w_1}\right)^{w_1} \left(\frac{u_2}{w_2}\right)^{w_2} \left(\frac{u_3}{w_3}\right)^{w_3} \left(\frac{u_4}{w_4}\right)^{Mw_4} \left(\frac{u_5}{w_5}\right)^{Mw_5} \tag{10-40}$$

Momentarily set aside the representations of Eqs. (10-33) to (10-40) and return to Eqs. (10-31) and (10-32), and solve by Lagrange multipliers

$$\nabla(u_1 + u_2 + u_3) - \lambda[\nabla(u_4 + u_5)] = 0 \tag{10-41}$$

$$u_4 + u_5 = 1 \tag{10-42}$$

The vector equation, Eq. (10-41), represents four scalar equations, the terms of which are the partial derivatives with respect to x_1 through x_4. Taking advantage of the fact that all of the u's are polynomials, each of the scalar equations of Eq. (10-41) can be multiplied by the variable with respect to which it has just been differentiated. The result is

$$a_{11}u_1^* + a_{21}u_2^* + a_{31}u_3^* - \lambda a_{41}u_4^* - \lambda a_{51}u_5^* = 0$$

$$\cdots \cdots \cdots \cdots \cdots \cdots \cdots \cdots \cdots \cdots \cdots \cdots \qquad (10\text{-}43)$$

$$a_{14}u_1^* + a_{24}u_2^* + a_{34}u_3^* - \lambda a_{44}u_4^* - \lambda a_{54}u_5^* = 0$$

The asterisk $*$ in Eqs. (10-43) indicates that these are optimal values of u. Next the results of Eqs. (10-43) can be merged with the representation of Eqs. (10-33) to (10-40). Dividing Eqs. (10-43) by y^* permits replacement of u_1^*/y^* by w_1 and similarly for u_2^*/y^* and u_3^*/y^*.

Since the constant M in Eq. (10-39) was arbitrary, let it equal $-\lambda/y^*$. The revised Eqs. (10-43), then, are

$$a_{11}w_1 + a_{21}w_2 + a_{31}w_3 + Ma_{41}w_4 + Ma_{51}w_5 = 0$$

$$\cdots \cdots \cdots \cdots \cdots \cdots \cdots \cdots \cdots \cdots \cdots \qquad (10\text{-}44)$$

$$a_{14}w_1 + a_{24}w_2 + a_{34}w_3 + Ma_{44}w_4 + Ma_{54}w_5 = 0$$

along with the conditions

$$w_1 + w_2 + w_3 = 1 \qquad\qquad (10\text{-}34)$$

and from Eq. (10-37)

$$Mw_4 + Mw_5 = M \qquad\qquad (10\text{-}45)$$

provide six linear, simultaneous equations. These six equations can be solved for the six unknowns w_1, w_2, w_3, Mw_4, Mw_5, and M.

It is especially significant that the equations in Eq. (10-44) require the cancellation of all the x terms in Eq. (10-40). This fact permits a convenient evaluation of the optimum value of y.

$$y^* = \left(\frac{c_1}{w_1}\right)^{w_1} \left(\frac{c_2}{w_2}\right)^{w_2} \left(\frac{c_3}{w_3}\right)^{w_3} \left(\frac{c_4}{w_4}\right)^{Mw_4} \left(\frac{c_5}{w_5}\right)^{Mw_5}$$

Example 10-5 Determine the minimum value of y where

$$y = \frac{40}{x_1 x_2 x_3} + 40 x_2 x_3$$

subject to the constraint

$$2x_1 x_3 + x_1 x_2 = 4$$

Solution Revise the constraint to

$$\frac{x_1 x_3}{2} + \frac{x_1 x_2}{4} = 1$$

The optimum value of y is

$$y^* = \left(\frac{40}{w_1}\right)^{w_1} \left(\frac{40}{w_2}\right)^{w_2} \left(\frac{\frac{1}{2}}{w_3}\right)^{Mw_3} \left(\frac{\frac{1}{4}}{w_4}\right)^{Mw_4}$$

provided that

$$-w_1 + Mw_3 + Mw_4 = 0$$
$$-w_1 + w_2 + Mw_4 = 0$$
$$-w_1 + w_2 + Mw_3 = 0$$
$$w_1 + w_2 = 1$$
$$Mw_3 + Mw_4 = M$$

Solving, $w_1 = \frac{2}{3}$, $w_2 = \frac{1}{3}$, $M = \frac{2}{3}$, $Mw_3 = \frac{1}{3}$, and $Mw_4 = \frac{1}{3}$, then

$$y^* = \left(\frac{40}{\frac{2}{3}}\right)^{\frac{2}{3}} \left(\frac{40}{\frac{1}{3}}\right)^{\frac{1}{3}} \left(\frac{\frac{1}{2}}{\frac{1}{2}}\right)^{\frac{1}{3}} \left(\frac{\frac{1}{4}}{\frac{1}{2}}\right)^{\frac{1}{3}} = 60$$

Finally, the individual values of x^* can be found from the distribution of the u terms. From Eq. (10-35),

$$u_1 = w_1 y^*$$

so

$$\frac{40}{x_1 x_2 x_3} = \frac{2}{3} y^* = 40$$

$$u_2 = w_2 y^*$$

so

$$40 x_2 x_3 = \frac{1}{3} y^* = 20$$

From Eq. (10-38),

$$u_3 = w_3 = \frac{1}{2} = \frac{x_1 x_3}{2}$$

$$u_4 = w_4 = \frac{1}{2} = \frac{x_1 x_2}{4}$$

So, $x_1^* = 2$, $x_2^* = 1$, and $x_3^* = \frac{1}{2}$.

10-9 Sensitivity coefficient

In Sec. 7-10 the Lagrange multiplier λ emerged as the sensitivity coefficient or dy^*/dA where A is the numerical term on the right side of the constraint equation. In the development of the procedure for constrained optimization using geometric programming the expression $-\lambda/y^*$ was replaced by the term M. When the optimization is complete and y^* and

M are known, the sensitivity coefficient can be determined quickly. In Example 10-5, $M = \frac{2}{3}$ and $y^* = 60$, so the sensitivity coefficient $\lambda = -40$. This value of the sensitivity coefficient is applicable to the constraint equation in the form

$$\frac{x_1 x_3}{2} + \frac{x_1 x_2}{4} = 1$$

For the constraint equation in the original form,

$$2x_1 x_3 + x_1 x_2 = 4$$

the sensitivity coefficient is $-40/4 = -10$.

10-10 Posynomials An inspection of the examples used in this chapter reveals the disturbing fact that the coefficients of all the terms in the objective function and constraints were positive. Duffin[1] called such equations posynomials to which the theory of geometric programming applies directly. While not explained in this chapter, it is possible to apply geometric programming to objective functions that have both positive and negative terms.[2]

10-11 Extensions of geometric programming In addition to extending geometric programming to functions with negative coefficients, it is possible to apply this method of optimization to problems where the constraints are inequalities which may have mixed senses.

The polynomial representation of the objective function to which geometric programming applies is frequently encountered because polynomials are a convenient form of mathematical modeling. Geometric programming is not applicable to such functions as logarithms, although in certain cases it may be advantageous to convert such functions to a polynomial series in the range of the expected solution in order to make geometric programming applicable.

It seems that geometric programming is a useful tool to carry in the optimization kit. There are certain situations (zero degrees of difficulty) where geometric programming is probably superior to all other optimization methods. When the degree of difficulty is greater than zero, it is questionable whether geometric programming can compete with other optimization methods.

---

[1] Ibid.
[2] D. G. Wilde and C. S. Beightler, "Foundations of Optimization," Prentice-Hall, Inc., Englewood Cliffs, N.J., 1967.

BIBLIOGRAPHY

Duffin, R. J., E. L. Peterson and C. M. Zener,: "Geometric Programming,"
John Wiley & Sons, Inc., New York, 1967.

Wilde, D. J., and C. S. Beightler: "Foundations of Optimization," Prentice-
Hall, Inc., Englewood Cliffs, N.J., 1967.

PROBLEMS

Solve the following problems by geometric programming.

10-1. The thickness of the insulation of a hot-water tank is to be selected
such that the total cost of the insulation and standby heating for the 10-year
life of the facility will be minimum.

Data:	Average water temperature	140°F
	Average ambient temperature	75°F
	Conductivity of insulation	0.02 Btu/(hr)(ft)(°F)
	Cost of heat	0.2 cents per thousand Btu
	Cost of insulation	$0.15(x)^{0.8}$ dollars/ft²

where x = thickness, in.
The operation is continuous. Assume that the only resistance to heat transfer
is the insulation.

(a) What is the minimum total cost of insulation and standby heat per
square foot of heat-transfer area for 10 years, neglecting interest charges?

(b) What is the optimum insulation thickness?

Ans.: $1.09, 5.62 in.

10-2. A hot-water boiler consists of a combustion chamber and a heat exchanger
arranged as shown in Fig. 10-3. Twenty pounds per hour of fuel having a
heating value of 18,000 Btu/lb burns in the combustion chamber. Thereafter,
the flue gases are cooled to 300°F while heating the water. The combustion
efficiency increases with an increase in the rate of air flow according to the
equation

$$\eta \, (\%) = 100 - \frac{800,000}{(m_a + 20)^2}$$

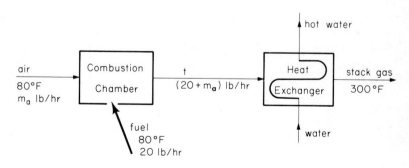

Fig. 10-3 Hot-water boiler in Prob. 10-2.

The specific heat of the gas mixtures is 0.25 Btu/(lb)(°F). Determine the value of m_a that results in the maximum rate of heat transfer to the water.

Ans.: 450 lb/hr.

10-3. A three-stage air compression system is being designed wherein the system receives air at 1 atm pressure absolute and 70°F and discharges the air at 64 atm absolute. Between each stage, the air passes through an intercooler which brings the temperature back to 70°F.

    (a) With the intermediate pressures chosen such that the total work of compression is a minimum, what is the minimum work required to compress 1 lb of air to the discharge pressure, assuming that the compressions are reversible and adiabatic?

    (b) What are the intermediate pressures that result in minimum work of compression?

Ans.: 184 Btu/lb, 4 and 16 atm.

10-4. The heat-rejection system for a condenser of a steam power plant, as shown in Fig. 10-4, is to be designed for minimum first-plus-pumping cost. The heat-rejection rate from the condenser is 4.8×10^7 Btu/hr. The costs in dollars that must be included are

    First cost of cooling tower $= 200A^{0.6}$

where $A = $ area, ft^2

    Lifetime pumping cost $= 0.5 \times 10^{-15}m^3$

where $m = $ flow rate, lb/hr

    Lifetime penalty in power production due to elevation of temperature of cooling water $= 100t$

where $t = $ temperature of water entering the condenser, °F

    The rate of heat transfer from the cooling tower can be represented adequately by the expression

    q (Btu/hr) $= 10^{-5}m^{1.2}tA$

    (a) Set up an unconstrained objective function in terms of variables A and m.

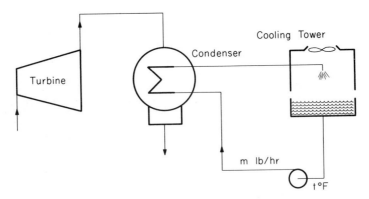

Fig. 10-4 Heat-rejection system in Prob. 10-4.

(b) Determine the minimum cost.

(c) Calculate the optimal values of A and m.

Ans.: $28,900, 1,440 ft^2, 1.96 × 10^6 lb/hr.

10-5. The first cost of a facility is $384,000 and the annual operating and maintenance costs increase with time according to the equation

$$4,000t + 750t^2 \qquad \$/\text{year}$$

where t = time in years from the start of operation

The life of the facility T years is sought such that the minimum average annual expense, neglecting interest, results. A long life results in skyrocketing operating and maintenance cost, while a short life spreads the first cost over only a small number of years.

(a) Write the objective function for the average annual cost in terms of T.

(b) Using geometric programming, solve for the minimum annual cost and the life that results in that minimum cost.

Ans.: $80,000, 8 years.

10-6. The total cost of a rectangular building shell and the land that it occupies is to be minimized for a building that must have a volume of 500,000 ft^3. The following costs per ft^2 apply:

Land	$8.00
Roof	1.25
Floor	0.75
Walls	1.00

Setting up the problem as one of constrained optimization, determine the minimum cost and the building dimensions at the optimum.

Ans.: $64,500, 46.5 × 46.5 × 232 ft.

10-7. Newly harvested grain often has a high moisture content and must be dried to prevent spoilage. This drying can be achieved by warming ambient air and blowing it through a bed of the grain. The seasonal operating cost in cents per square foot of grain bed for such a dryer consists of the cost of heating the air.

$$\text{Heating cost (cents/ft}^2) = 0.004Q \, (\Delta t)$$

and the blower operating cost is

$$\text{Blower cost (cents/ft}^2) = (1 \times 10^{-7})Q^3$$

where Q = air quantity delivered through the bed during the season in thousands of ft^3 (m-ft^3)

Δt = the rise in air temperature during heating, °F

The values of Q and Δt also influence the time required for adequate drying of the grain according to the equation

$$\text{Drying time (days)} = \frac{7,200,000}{(Q^2)(\Delta t)}$$

Using the geometric-programming method of constrained optimization, compute the minimum operating cost and the optimum values of Q and Δt that will achieve adequate drying in 60 days.

Ans.: 3.2 cents, 200 m-ft^3, 3°F.

11
Linear Programming

11-1 Introduction Linear programming is an optimization method applicable where both the objective function and the constraints can be expressed as linear combinations of the variables. The constraint equations may be equalities or inequalities. Linear programming first appeared in Europe in the 1930s when economists and mathematicians began working on economic models. During World War II the Air Force sought more effective procedures of allocating resources and turned to linear programming. A member of the group working on the Air Force problem was George Dantzig who in 1947 reported the *simplex method* for linear programming which was a significant step in bringing linear programming into wider usage.

Economists and industrial engineers have applied linear programming more in their fields of work than have most other technical groups. Decisions regarding time allocations of machines to various products in a manufacturing plant, for example, are ones that lend themselves neatly to linear programming. Linear programming has not so far played an important role in thermal systems with the exception of petroleum processing. Most large oil companies use linear programming models to determine what quantities and sources of crude oil should be purchased and the quantities of the various products that will result in optimum profit for the entire operation. Within the refinery itself, linear programming aids in determining where the bottlenecks to production exist, and how much the total output of the plant could be increased, for example, by enlarging a heat exchanger, or by increasing a certain flow rate.

A thorough study of linear programming requires a knowledge of matrix algebra, but the short explanation in this chapter can serve as an introduction to the subject. The emphasis is on obtaining a geometric feel of the linear-programming situation, although the simplex algorithm for solving both maximization and minimization problems in linear programming will be explained and used for several straightforward situations.

11-2 Some examples of linear programming Some classic uses of linear programming are to solve (1) the blending problem, (2) machine allocation, (3) inventory and production planning, and (4) the transportation problem.

The application of the oil company mentioned in Sec. 11-1 is typical of the blending application. The oil company has a choice of buying crude from several different sources with differing compositions and at differing prices. It has a choice of manufacturing various quantities of aviation fuel, automobile gasoline, diesel fuel, and oil for heating. The combinations of these products are restricted by material balances based on the incoming crude and by the capacity of such components in the refinery as the cracking unit. A mix of purchased crude and manufactured products is sought that gives maximum profit.

The machine allocation problem occurs where a manufacturing plant has a choice of making several different products, each of which requires varying machine times of different machines such as lathes, screw machines, and grinders. The machine time of some or all of these different machines is limited. The goal of the analysis is to determine the production quantities of each product that result in maximum profit.

The sales of some manufacturing firms fluctuate, often according to a seasonal pattern. The company can build up an inventory of manufactured products to carry it through the period of peak sales, but carrying an inventory costs money. Or it can pay overtime rates in order to step up its production during the period of peak sales, which also entails an additional expense. Finally, the company can simply plan on losing some sales because it does not meet the sales demand at the time that it exists, thus losing a potential profit. Linear programming can incorporate the various cost and loss factors and arrive at the most profitable production plan.

The fourth application, the transportation problem, occurs where an organization has several production plants distributed throughout a geographical area and also has a number of warehouses placed in other geographic locations. Each plant has a certain production capability and each warehouse has a certain requirement. Any one warehouse may receive the production from one or more plants. The object is to determine how much of each plant's production should be shipped to each warehouse in order to minimize the total manufacturing and transportation cost.

Simple linear-programming problems may be done in a hit-or-miss fashion, but the ones with more than two or three variables require systematic procedures. Even when using methodical techniques, the magnitude of a problem that can be solved by hand is limited. Large problems—ones with several thousand variables are now being solved—require computer programs, which are currently available as library routines.

11-3   Mathematical statement of the linear-programming problem

The form of the statement is typical of the optimization problem in that it consists of an objective function and constraints. The objective function which is to be minimized (or maximized) is

$$y = c_1 x_1 + c_2 x_2 + \cdots + c_n x_n \tag{11-1}$$

and the constraints

$$\phi_1 = a_{11} x_1 + a_{12} x_2 + \cdots + a_{1n} x_n \geq r_1$$
$$\cdots \cdots \cdots \cdots \cdots \cdots \cdots \cdots \cdots \cdots \cdots$$
$$\phi_m = a_{m1} x_1 + a_{m2} x_2 + \cdots + a_{mn} x_n \geq r_m \tag{11-2}$$

Furthermore, if the x's represent physical quantities, they are likely to be nonnegative, so $x_1, \ldots, x_n \geq 0$. The c values and a values are all constants which make both the objective function and the constraints linear, and thus the name, *linear programming*. The values of "c" and "a" may be either positive, negative, or zero. The inequalities in the constraints can be in either direction and can even be strict equalities. In this chapter, however, we shall consider only equality-inequality possibilities, such as those shown in Eq. (11-2) where the sense of all inequalities is the same.

At first glance, this problem might seem readily soluble by Lagrange multipliers, but we recall that the method of Lagrange multipliers is applicable where *equality* constraints exist. Furthermore, Lagrange multipliers apply where $n > m$, but in linear programming n can be greater, equal, or less than m. The significance of $n < m$ will be discussed in Sec. 11-14.

11-4 Developing the mathematical statement The translation of the physical conditions into a mathematical statement of linear-programming form will be illustrated by an example.

Example 11-1 A simple power plant consists of an extraction turbine that drives a generator, as shown in Fig. 11-1. The turbine receives 25,000 lb/hr (25 m-lb/hr) of steam and the plant can sell either electricity or extraction steam for processing purposes. The revenue rates are:

Electricity	$0.02 per kWh
Low-pressure steam	0.34 per m-lb
High-pressure steam	0.50 per m-lb

The generation rate of electrical power depends upon the flow rate of steam passing through each of the sections A, B, and C, and these flow rates are w_A, w_B, and w_C m-lb/hr, respectively. The relationships are:

$$P_A \text{ (kW)} = 6(w_A)$$
$$P_B \text{ (kW)} = 7(w_B)$$
$$P_C \text{ (kW)} = 10(w_C)$$

The plant can sell as much electricity as it generates, but there are other restrictions:

To prevent overheating of the low-pressure section of the turbine, no

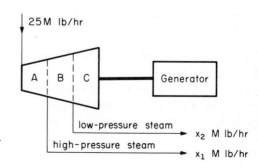

Fig. 11-1 Power plant in Example 11-1.

less than 5 m-lb/hr must always flow through section C. Furthermore, to prevent unequal loading on the shaft, the permissible combination of extraction rates is such that if $x_1 = 0$ that $x_2 \leq 15$ m-lb/hr, and for each pound of x_1 extracted $\frac{1}{4}$ lb less can be extracted of x_2.

The customer of the process steam is primarily interested in total Btu and will purchase no more than given by the equation

$$4x_1 + 3x_2 \leq 72$$

Develop the objective function for the total revenue from the plant and also the constraint equations.

Solution   The revenue y is the sum of the individual revenues from x_1, x_2, and the electric power.

$$y = 0.50x_1 + 0.34x_2 + (0.02)(6w_A + 7w_B + 10w_C)$$

Since, from a mass balance, $w_A = 25$, $w_B = 25 - x_1$, and

$$w_C = 25 - x_1 - x_2$$
$$y = 0.50x_1 + 0.34x_2 + (0.12)(25) + (0.14)(25 - x_1)$$
$$+ (0.20)(25 - x_1 - x_2)$$

and

$$y = 11.50 + 0.16x_1 + 0.14x_2$$

Because the constant has no effect on the state point at which the optimum occurs, the objective function to be maximized is:

$$y = 0.16x_1 + 0.14x_2 \tag{11-3}$$

The three constraints are as follows:

$$x_1 + x_2 \leq 20 \tag{11-4}$$
$$x_1 + 4x_2 \leq 60 \tag{11-5}$$
$$4x_1 + 3x_2 \leq 72 \tag{11-6}$$

11-5 Geometric visualization of the linear–programming problem

Example 11-1, since it involves only the two variables x_1 and x_2, can be illustrated geometrically as in Fig. 11-2. The constraint of Eq. (11-4), for example, states that only the region on and to the left of the line $x_1 + x_2 = 20$ is permitted. Placing the other two constraints in Fig. 11-2 further restricts the permitted region to A-B-D-F-G.

Next the lines of constant revenue y are plotted on Fig. 11-2. Inspection shows that the greatest profit can be achieved by moving to point D where $x_1 = 12$ and $x_2 = 8$. An important generalization is that the *optimum solution lies at a corner*. A special case of this generalization is where the line of constant profit is parallel

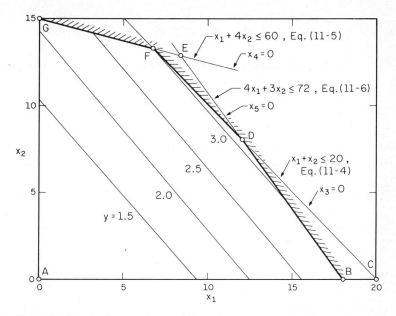

Fig. 11-2　Graph of constraints and lines of constant profit in Example 11-1.

to a constraint line, in which case any point on the constraint line between the corners is equally favorable.

If the objective function depends upon three variables, a three-dimensional graph is required, and then the constraint equations are represented by planes. The corner where the optimum occurs is formed by the intersection of three planes.

11-6　Introduction of slack variables

The constraint equations, Eqs. (11-4) to (11-6), are inequalities, but they can be converted to equalities by the introduction of another variable in each equation.

$$x_1 + x_2 + x_3 \qquad\qquad = 20 \qquad\qquad (11\text{-}7)$$

$$x_1 + 4x_2 \qquad + x_4 \qquad = 60 \qquad\qquad (11\text{-}8)$$

$$4x_1 + 3x_2 \qquad\qquad + x_5 = 72 \qquad\qquad (11\text{-}9)$$

This substitution is valid provided that $x_3 \geq 0$, $x_4 \geq 0$, $x_5 \geq 0$. These new variables are called *slack variables*.

Reference to Fig. 11-2 permits a geometric interpretation of the slack variables x_3, x_4, and x_5. Any point on the graph defines specific values of x_3, x_4, and x_5. Along the $x_1 + x_2 = 20$ line, for example, $x_3 = 0$. The value of x_3 in the region to the right of the line is less than zero and is thus prohibited, while on the line and to the left of it $x_3 \geq 0$, and this region is permitted.

11-7 Determining optimum values by solving simultaneous equations

The set of equations, Eqs. (11-7) to (11-9), cannot be solved for unique values of the unknowns, because there are more unknowns than equations. It would be possible to solve for three of the variables in terms of the other two. Another form of the same statement would be to say that two of the variables could be assigned arbitrary values and numerical answers could be obtained for the other three variables. A useful choice of these two arbitrary values would be zero, because we recall from Sec. 11-5 that the optimum lies at a corner, and a corner is formed by the intersection of two lines along which two corresponding variables are zero.

The number of possible combinations of solutions of the set, Eqs. (11-7) to (11-9), where two variables are equated to zero is 10. Of these 10 solutions, 7 are shown in Fig. 11-2 as points A through G. The summary of these solutions is shown in Table 11-1. The revenue calculated from the objective function, Eq. (11-3), is also tabulated.

Not all of the solutions are permitted, because some of the intersections occur in the prohibited region. This condition is indicated by the slack variable going negative, and the solution must be excluded since a constraint equation is not satisfied. These permitted solutions are called *feasible solutions*. An *infeasible solution* violates a constraint.

Of the feasible solutions, point D where $x_1 = 12$ and $x_2 = 8$

Table 11-1 Simultaneous solutions of Eqs. (11-7) to (11-9). Variables arbitrarily assigned zero are designated by *

Point on Fig. 11-2	x_1	x_2	x_3	x_4	x_5	Revenue y	Solution excluded
A	*	*	20.0	60.0	72.0	0.00	
	*	20.0	*	−20.0	12.0	2.80	×
G	*	15.0	5.0	*	27.0	2.10	
	*	24.0	−4.0	−36.0	*	3.36	×
C	20.0	*	*	40.0	−8.0	3.20	×
	60.0	*	−40.0	*	−168.0	9.60	×
B	18.0	*	2.0	42.0	*	2.88	
F	6.67	13.33	*	*	5.32	2.94	
D	12.0	8.0	*	16.0	*	3.04	
E	8.32	12.92	−1.24	*	*	3.14	×

provides the maximum revenue of 3.04. At this point $x_4 = 16$, which indicates that constraint equation (11-5) plays no role in the solution. If it were suggested that some expenditure be made to relax this constraint, this expenditure would be fruitless. On the other hand, relaxing the constraints of Eqs. (11-4) and (11-6) would be productive in increasing the value of the optimum revenue.

Showing the situation of the simultaneous equations as a means of attacking the linear-programming problem gives further insight into the nature of the problem and the solution. It is not, however, the recommended method of solution. Even in this small problem, it was necessary to solve three-equation sets ten times! The power of the *simplex algorithm* that will be introduced next is that only a small number of the total number of points indicated in Table 11-1 are ever examined. The infeasible solutions and less attractive points are ignored.

11-8 Introduction to the simplex method for a maximization problem

To lead gradually into the simplex method, an exploded view of the process will first be portrayed. The procedure will be essentially the same as the simplex algorithm, but will lack the efficiency and neatness of the simplex algorithm. The explanation will be applicable to a maximization problem where the inequality constraints have a less-than sense, such as is true of Example 11-1. Geometrically, the process is one of starting at a feasible point, and then moving in the feasible region to the adjacent corner that gives the best improvement in the objective function. This process continues until no further improvement is possible.

The steps in the exploded view are as follows:

1. Write the constraint equations including the slack variables.
2. Denote which x's are zero.
3. State the objective function and calculate its value.
4. Decide which x to change.
5. Determine which equation controls the change of the variable in step 4.
6. From the controlling equation of step 5, solve for the x being changed and substitute this expression into new steps 1 and 3.

This method will now be used to solve Example 11-1. The stages of the solution, each of which contain the six steps listed above, will be designated stage I, stage II, etc.

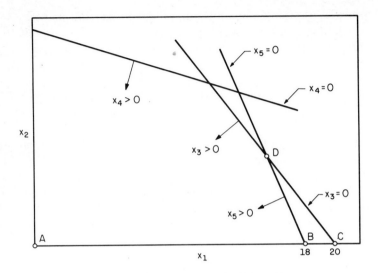

Fig. 11-3 Moving around the corners to find the optimum.

Stage I

1. The constraint equations are the original ones, Eqs. (11-7), (11-8), and (11-9),

$$x_1 + x_2 + x_3 \qquad\qquad = 20$$
$$x_1 + 4x_2 \qquad + x_4 \qquad = 60$$
$$4x_1 + 3x_2 \qquad\qquad + x_5 = 72$$

2. In order to denote which x's are zero for the first stage, we must decide upon the initial feasible point. In a complex problem, this decision may be difficult, but in the maximization problem with less-than constraints, the simple choice is to set the physical variables equal to zero. This is a feasible solution since the constraint equations in step 1 are satisfied. Therefore, $x_1 = 0$, $x_2 = 0$, and from the constraint equations $x_3 = 20$, $x_4 = 60$, and $x_5 = 72$. In Fig. 11-3 this first feasible point is at the origin, point A.

3. The objective function is the original one

$$y = 0.16x_1 + 0.14x_2$$

and at this first point, since $x_1 = x_2 = 0$, $y = 0$.

4. The statement of the objective function in step 3 indicates which x variable should be changed. A decrease in either x_1 or x_2

would force the variable negative, which is not permitted. Furthermore, we see that an increase in either x_1 or x_2 is needed to increase y. The coefficient of x_1 is 0.16 which is higher than the coefficient of x_2, so a given increase in x_1 is more effective in increasing y than is true of x_2. The procedure then, is to increase x_1, meanwhile holding $x_2 = 0$.

5. The next task is to determine to what extent x_1 can be increased. The constraint equations in step 1 decide this question. Since x_2 is zero, the first equation indicates that x_1 can increase to a value of 20. An increase beyond 20 forces the slack variable x_3 negative which is our way of showing infeasibility. The limiting values of x_1 demanded by the three constraints are:

First constraint: $x_1 = 20$

Second constraint: $x_1 = 60$

Third constraint: $x_1 = 18$

so the third constraint, which is the most restrictive, is the one that controls. In Fig. 11-3, the foregoing process was one of moving out along the x_1 axis to the intersection with the first constraint line. We have found that we can move no further than to point B where $x_1 = 18$, which is where the slack variable $x_5 = 0$.

6. This next step should be examined deliberately, because it is a conceptual feature of linear programming. We accept the fact that a point can be located in Fig. 11-3 by designating the values of x_1 and x_2. It is also possible to fix a point where one or both of the independent variables are slack variables. For example, point B can be located by specifying $x_2 = 0$ and $x_5 = 0$. Furthermore, recalling that the entire linear–programming process is to be one of moving around corners, we would like to move along the $x_5 = 0$ line to the next corner. The way in which we can be sure of moving along this line is to introduce the x_5 variable in our constraint equations and objective function. This operation is accomplished by solving for x_1 in the controlling constraint equation and substituting this expression into the remaining constraint equations and the objective function.

From the third constraint

$$x_1 = 18 - 0.25x_5 - 0.75x_2$$

This completes the steps of stage I.

Stage II

1. The new constraint equations are formed by substituting the x_1 expression into the former constraint equations. The revised first constraint is

$$(18 - 0.25x_5 - 0.75x_2) + x_2 + x_3 = 20$$

or

$$0.25x_2 + x_3 - 0.25x_5 = 2$$

The second and third constraints are

$$3.25x_2 + x_4 - 0.25x_5 = 42$$
$$x_1 + 0.75x_2 + 0.25x_5 = 18$$

The controlling equation was divided through by the coefficient of the x_1 term for future convenience.

2. The x's that are zero are x_2 which has remained zero so far, and x_5 which we hold at zero in order to move along the constraint that it designates. With $x_2 = x_5 = 0$, the constraint equations readily yield the values of the other x's: $x_3 = 2$, $x_4 = 42$, and $x_1 = 18$.

   A pattern that can be seen developing is that at each point to which the process moves, two variables (the same number as physical variables) are equal to zero. Then one of these zero variables is increased until stopped by a constraint.

3. The new objective function contains only x_2 and x_5 after the substitution of the expression for x_1 from step 6 of stage I,

$$y = 2.88 - 0.04x_5 + 0.02x_2$$

The value of y at point B is immediately found to be 2.88 by substituting $x_2 = x_5 = 0$.

4. At this point, the decision is made about the next move. Examination of the objective function shows that y cannot be increased by increasing x_5, and x_5 cannot be decreased because it is already zero and a decrease would force it negative which violates a constraint. If x_5 is increased the value of y decreases because of the negative coefficient of x_5. Geometrically, increasing x_5 amounts to moving toward the left of the x_5-constraint line in Fig. 11-3. The variable x_2 offers more profitable possibilities, because an increase in x_2 increases y.

5. Holding $x_5 = 0$, the constraint equations listed in step 1 of this stage show:

First constraint: $x_2 = 8$ when x_3 reduces to zero

Second constraint: $x_2 = 12.95$ when x_4 reduces to zero

Third constraint: $x_2 = 24$ when x_1 reduces to zero

6. The first constraint controls, indicating that the point has moved to where $x_3 = 0$. Using this equation to solve for x_2,

$$x_2 = 8 - 4x_3 + x_5$$

Stage III

1. Substituting the expression for x_2 and dividing the controlling equation by the coefficient of x_2 yields

$$x_2 + 4x_3 \qquad - x_5 = 8$$
$$- 13x_3 + x_4 + 3x_5 = 16$$
$$x_1 \quad - 3x_3 \qquad + x_5 = 12$$

2. At the beginning of stage II the zero variables were x_2 and x_5. In the programming of stage II, x_2 increased until x_3 reduced to zero, meanwhile x_5 remained zero. So $x_3 = x_5 = 0$, and when those values are substituted into the constraint equations, the other variables are found.

$$x_2 = 8 \qquad x_4 = 16 \qquad x_1 = 12$$

3. The new form of the objective function, after substituting the expression for x_2, is

$$y = 3.04 - 0.08x_3 - 0.02x_5$$

At the current location, which is point D, the value of y is 3.04.

4. Considering the next move, we rule out decreases in x_3 and x_5 because they violate constraints. But increases in either x_3 or x_5 result in a reduction of y. We conclude, then, that there is no further increase possible in y, and we have found the maximum at

$$x_1 = 12 \qquad x_2 = 8 \qquad y = 3.04$$

Reviewing the procedure, the progression was from point A to B to D in Fig. 11-3. The objective function increased in value with each move from 0 to 2.88 to the maximum of 3.04. At each point two of the five variables were zero, and the translation from one point to the next consisted of replacing one of the zero variables with a different one. The coefficients of the variables in the successively revised objective function gave the clue as to which zero variable to increase.

11-9 Basic variables Purely as a matter of terminology, the non-zero variables at each stage are called *basic variables*. In the three stages in the solution of Example 11-1, the basic variables are marked by an asterisk:

	x_1	x_2	x_3	x_4	x_5
Stage I			*	*	*
Stage II	*		*	*	
Stage III	*	*		*	

In describing which are the basic variables, it is common to say, for example at stage I, that the "variables in the base are x_3, x_4, and x_5."

11-10 Presenting the constraint equations in tableaux This section and the several that follow convert the expanded procedure of Sec. 11-8 into a more systematic and efficient form. One of the features of the streamlined process will be to display the constraint equations and objective function in tableau form as they are successively revised throughout the solution. The other major feature is the simplex algorithm to methodically convert the current tableau into the next one.

After performing one or two solutions by the expanded procedure, it would probably occur to the worker that a table could be constructed that would eliminate the need of writing the x's in each equation. The original constraint equations could be written in such a form that only the coefficients and constant term appear in the body of the table. The double vertical line at the right represents the equality sign.

Partial Tableau 1   Example 11-1

	$x_1 = 0$	$x_2 = 0$	x_3	x_4	x_5	
$2 0 / 1 = 20$	①	1	1			20
$6 0 / 1 = 60$	1	4		1		60
$7 2 / 4 = 18$	4	3			①	72

(handwritten annotations: "limiting values", "Controls", "$x_5 = 0$ IN NEXT TABLEAU")

The fact that $x_1 = x_2 = 0$ at the first feasible point is indicated in the column headings. The next decision is to determine which of the zero variables to increase. This decision, which was dictated by the coefficients in the objective function, specifies that x_1 should be programmed because the coefficient of x_1 is a larger positive number than the coefficient of x_2. The vertical arrow in the x_1 column denotes this choice. Next to be determined are the limits to which x_1 can be increased, which are shown in the left margin of the tableau. Since $x_1 = 18$ is the most restrictive, the third constraint equation controls and x_5 will equal zero in the next tableau.

11-11 The simplex algorithm The next step in the expanded procedure is to solve for the variable being programmed (x_1 in this case) and substitute the expression into the other constraints and the objective function. The algebra becomes cumbersome if the equations contain a large number of variables. The simplex algorithm permits simpler, bite-sized steps by concentrating on just one position at a time in the new tableau. The procedure in calculating the new tableau is as follows:

1. Divide the controlling equation by the coefficient of the x variable being programmed and write the result in the new tableau.

For the rows representing the other constraints:

2. Select a box in the new tableau. Call the value in the same box of the old tableau v.
3. Move sideways in the old tableau to the x variable being programmed. Call this value w.
4. In the new tableau, move from the box being calculated up or down to the row which contains the previous controlling equation. Call the value in that box z.
5. The value of the box in the new tableau is $v - wz$.

Applying the above procedure to transform from the first to the second tableau in Example 11-1, the third or controlling equation is divided by 4 and rewritten in the new tableau.

Next calculate the value in each box, starting with the first row, first column. The value in the box in the old tableau is 1, so $v = 1$. The value of w is found by moving sideways in the old tableau to the variable being programmed, but that variable is x_1 where we are already located, so $w = 1$. Moving vertically in the new tableau to the previous controlling equation, we find $z = 1$. The value in the box of the new tableau is $v - wz$, so its value is $1 - (1)(1) = 0$.

Moving to the box in the first row, second column, $v = 1$, $w = 1$, and $z = 0.75$. The new value, then, is $1 - (1)(0.75) = 0.25$. These values check with those determined in the expanded procedure of Sec. 11-8.

All of the other boxes are calculated in a similar manner, including the numerical constant to the right of the double line, resulting in the following tableau:

Partial Tableau 2   Example 11-1

x_1	$x_2 = 0$	x_3	x_4	$x_5 = 0$	
0	0.25	1	0	-0.25	2
0	3.25	0	1	-0.25	42
1	0.75	0	0	0.25	18

11-12 Inclusion of the objective function in the tableau Before proceeding to the next tableau, we shall return to Tableau 1, pick up the objective function, and carry it along with the developing tableaus. The objective function and its successive transformations served two purposes in the expanded procedure of Sec. 11-8. The numerical constant in the equation gave a running account of the value of the objective function: 0 to 2.88 to 3.04. The other purpose was that the coefficients of the x variables indicated which variable should be programmed next. It is advantageous to carry along the objective function in the tableaux that are being developed, particularly since terms in the objective function submit themselves to the simplex algorithm.

The original objective function

$$y = 0.16x_1 + 0.14x_2$$

can be rewritten

$$y - 0.16x_1 - 0.14x_2 = 0$$

Placing the coefficients of the x's in the appropriate columns of a new fourth row in the first tableau, the zero lodges to the right of the equal sign represented by the double vertical line on the tableau. The interpretation of the fourth row is that since either the variable or the coefficient on the left side of the double line is zero, all that remains on the left side is the understood appearance of y which equals the value on the right side of the double line.

Tableau 1 Example 11-1

⇓

	$x_1 = 0$	$x_2 = 0$	x_3	x_4	x_5	
$2\%_1 = 20$	1	1	1			20
$6\%_1 = 60$	1	4		1		60
$7\frac{2}{4} = 18$	4	3			1	72
Difference coefficients	-0.16	-0.14				0

The numbers in the boxes in the fourth row are called *difference coefficients* and are the indicators of which variable to program next. In contrast to the way in which they appeared in the original objective function, their signs are reversed, so now the *largest negative* coefficient specifies the variable to be programmed next.

In moving to Tableau 2, the identical simplex algorithm that applies to the constraint equations applies equally well to the terms in the objective function equation.

Tableau 2 Example 11-1

⇓

$x_3 = 0$ IN NEXT TABLEAU

	x_1	$x_2 = 0$	x_3	x_4	$x_5 = 0$	
$\frac{2}{0}.25 = 8$	0	0.25	1	0	-0.25	2.0
$4\frac{2}{3}.25 = 12.95$	0	3.25	0	1	-0.25	42.0
$1\frac{8}{0}.75 = 24$	1	0.75	0	0	0.25	18.0
	0	-0.02	0	0	0.04	2.88

After completing the simplex operation to derive Tableau 2, examination of the difference coefficients shows that the largest negative value is the one associated with x_2, so x_2 is programmed until restricted by the first constraint. The third tableau becomes:

Tableau 3   Example 11-1

x_1	x_2	$x_3 = 0$	x_4	$x_5 = 0$	
	1	4		-1	8
		-13	1	3	16
1		-3		1	12
		0.08		0.02	3.04

There are no negative difference coefficients in Tableau 3, so there is no further possibility of increasing the objective function. Tableau 3 contains a statement of the optimal conditions: $x_1 = 12$, $x_2 = 8$, $x_3 = 0$, $x_4 = 16$, $x_5 = 0$, and $y = 3.04$.

11-13   Geometrical interpretation of tableau transformation   The displacement of one variable in the base by a new entering variable occurs during the transformation from one tableau to the next. Geometrically this transformation results in a change of coordinates and the appearance of constraint lines on the graph. We continue with the numerical values of Example 11-1. Figure 11-2 displays the equations of Tableau 1. The position in Fig. 11-2 at the starting point is the origin where $x_1 = x_2 = 0$, $y = 0$, $x_3 = 20$, $x_4 = 60$, and $x_5 = 72$.

In the transformation to Tableau 2, x_1 increases until x_5 reduces to zero. The equations of Tableau 2 can be shown on a graph such as Fig. 11-4 where now the coordinates are x_2 and x_5. The first equation in the tableau with $x_3 = 0$ is

$$0.25x_2 - 0.25x_5 = 2$$

To the right of the $x_3 = 0$ line is the region where x_3 is positive. The other two constraint lines—where $x_1 = 0$ and $x_4 = 0$—are also shown in Fig. 11-4. The dashed lines show lines of constant values of the objective function. The point described by Tableau 2 is the origin where $x_2 = x_5 = 0$ and $y = 2.88$. The geometry of Fig.

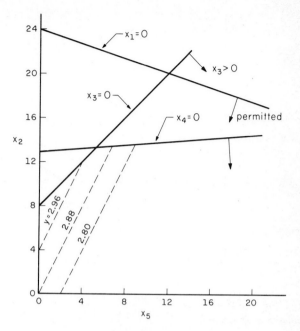

Fig. 11-4 Graph of Tableau 2.

11-4 confirms that an increase in x_5 decreases y, but an increase in x_2 causes a favorable change in y, so x_2 is the next variable brought into the base. The variable x_2 increases until its value is 8 at which point the $x_3 = 0$ constraint prohibits further increase in x_2.

Figure 11-5 shows the graphic representation of Tableau 3. Within the feasible region there is no further increase in y possible than that which exists at the origin where $y = 3.04$, so the optimum has been reached.

11-14 Number of variables and number of constraints The relationship of the number of physical variables and the number of constraints gives an indication of the number of variables that are zero in the solution. Let the number of physical variables be denoted by n, and the number of constraints and, therefore, the number of slack variables be denoted by m. There will always be n variables (physical plus slack) equal to zero at the optimum, or for that matter at any corner. When $m > n$, as Fig. 11-6a, at least $m - n$ constraints play no role in the solution. In Fig. 11-6b where $m < n$, at least $n - m$ physical variables are zero.

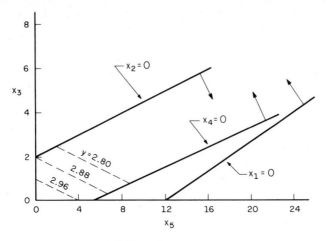

Fig. 11-5   Graph of Tableau 3.

11-15   Minimization   Solution of the maximization problem with less-than constraints consisted of moving from one corner in the feasible region to whichever adjacent corner showed the most improvement in the objective function.   Finding the first feasible point posed no problem because the origin, which was in the feasible region, was selected.   In the minimization problem with greater-than constraints, locating the first feasible point may be difficult. Admittedly, in simple problems involving a small number of variables, combinations of variables could be set to zero in the constraint equations and the other variables solved as was done in Sec. 11-7, until a combination is found that violates no constraints.

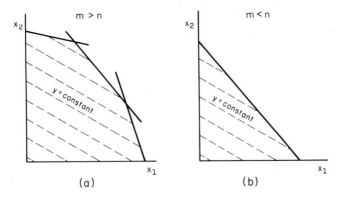

Fig. 11-6   Relation of number of physical and slack variables.

In large problems, this method is prohibitive and a more systematic procedure must be employed. Such a procedure, whose feature is the introduction of "artificial variables," will be illustrated in the following example.

Example 11-2 Determine the minimum value of y and the values of x_1 and x_2 at this minimum, where

$$y = 6x_1 + 3x_2$$

subject to the constraints

$$5x_1 + \quad x_2 \geq 10$$
$$9x_1 + 13x_2 \geq 74$$
$$x_1 + \quad 3x_2 \geq 9$$

Solution Since this problem involves only two physical variables, the constraints and lines of constant y can be graphed, as in Fig. 11-7.

For the solution by linear programming, first write the constraint inequalities as equations by introducing the slack variables x_3, x_4, and x_5.

$$5x_1 + \quad x_2 - x_3 \qquad\qquad = 10$$
$$9x_1 + 13x_2 \qquad - x_4 \qquad = 74$$
$$x_1 + \quad 3x_2 \qquad\qquad - x_5 = 9$$

The slack variables bear negative signs when the constraints have a

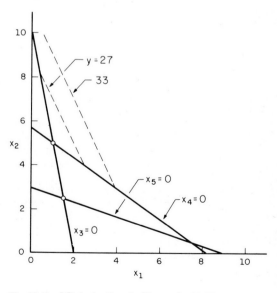

Fig. 11-7 Minimization in Example 11-2.

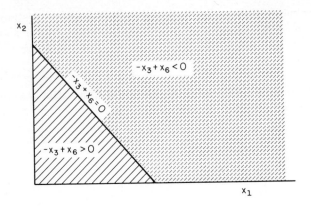

Fig. 11-8   Relationship of slack and artificial variables.

greater-than sense in order that the slack variable be equal to or greater than zero when the constraint is satisfied.

Next comes the unique feature of the minimization problem—the introduction of artificial variables.   An artificial variable is introduced for each slack variable in the constraint equations

$$5x_1 + \quad x_2 - x_3 \qquad\qquad + x_6 \qquad\qquad = 10 \qquad\qquad (11\text{-}10)$$

$$9x_1 + 13x_2 \qquad - x_4 \qquad\qquad + x_7 \qquad = 74 \qquad\qquad (11\text{-}11)$$

$$x_1 + \quad 3x_2 \qquad\qquad - x_5 \qquad\qquad + x_8 = 9 \qquad\qquad (11\text{-}12)$$

The objective function is also revised to incorporate the artificial variables

$$y = 6x_1 + 3x_2 + Px_6 + Px_7 + Px_8 \qquad\qquad (11\text{-}13)$$

The coefficient P of the artificial variable never assumes a numerical value, but is only considered to be extremely large.   The existence of the product of P and the artificial variables in the objective function is so penalizing to the minimization attempt that no satisfactory minimization will occur until the values of the artificial variables are reduced to zero.

Before proceeding with the solution, an examination of the effect of the artificial variables would be helpful.   Isolating just one of the constraints—Eq. (11-10)—and showing it in Fig. 11-8, the combination $-x_3 + x_6 = 0$ applies along the constraint line.   Below and to the left of the constraint line, $-x_3 + x_6 > 0$.   It is possible to move into this region by holding x_3 equal to zero and increasing x_6.   Such a decision has the associated penalty of making the objective function very large, but it does convert the region into a feasible one in the sense that no variables are negative.

The geometry of Fig. 11-8 suggests that the origin can once again be chosen as the starting point.   In fact, all the physical and slack variables can be set equal to zero, $x_1 = x_2 = x_3 = x_4 = x_5 = 0$, by permitting x_6, x_7, and x_8 to take on positive values.

The simplex algorithm applies, but one further operation is performed on the objective function before writing the first tableau. From Eqs. (11-10) to (11-12),

$$x_6 = 10 - 5x_1 - x_2 + x_3$$
$$x_7 = 74 - 9x_1 - 13x_2 + x_4$$
$$x_8 = 9 - x_1 - 3x_2 + x_5$$

Substituting these values of the artificial variables into Eq. (11-13) and grouping,

$$y = (6 - 15P)x_1 + (3 - 17P)x_2 + Px_3 + Px_4 + Px_5 + 93P$$

The first tableau can now be constructed. In the minimization operation the variable with the *largest positive* difference coefficient is chosen to be programmed. In Tableau 1 this choice will be x_2 because P appears with the largest coefficient and the P values dominate over any purely numerical terms.

Tableau 1   Example 11-2

⇓

	$x_1 = 0$	$x_2 = 0$	$x_3 = 0$	$x_4 = 0$	$x_5 = 0$	x_6	x_7	x_8	
10	5	1	-1			1			10
$7\frac{4}{13}$	9	13		-1			1		74
⇒ 3	1	3			-1			1	9
	$15P - 6$	$17P - 3$	$-P$	$-P$	$-P$				$93P$

The next question is, "What is the limit to which x_2 can be increased?" Just as in the maximization process the variable x_2 is increased to the most limiting constraint, which in this case is the third one. The procedure in transforming to Tableau 2 is the standard simplex algorithm.

Tableau 2   Example 11-2

⇓

	$x_1 = 0$	x_2	$x_3 = 0$	$x_4 = 0$	$x_5 = 0$	x_6	x_7	$x_8 = 0$	
⇒ $3\frac{1}{2}$	$1\frac{4}{3}$	0	-1	0	$\frac{1}{3}$	1	0	$-\frac{1}{3}$	7
$10\frac{5}{14}$	$1\frac{4}{3}$	0	0	-1	$13\frac{4}{3}$	0	1	$-13\frac{4}{3}$	35
9	$\frac{1}{3}$	1	0	0	$-\frac{1}{3}$	0	0	$\frac{1}{3}$	3
	$\dfrac{28P - 15}{3}$	0	$-P$	$-P$	$\dfrac{14P - 3}{3}$	0	0	$\dfrac{-17P + 3}{3}$	$42P + 9$

The difference coefficient with the largest positive value is $(28P - 15)/3$, so x_1 is programmed next until limited by the point where x_6 starts to go negative to the first constraint.

Tableau 3   Example 11-2

$\Downarrow$

	x_1	x_2	$x_3 = 0$	$x_4 = 0$	$x_5 = 0$	$x_6 = 0$	x_7	$x_8 = 0$	
21	1	0	$-3/14$	0	$1/14$	$3/14$	0	$-1/14$	$3/2$
7	0	0	1	-1	4	-1	1	-4	28
-7	0	1	$1/14$	0	$-5/14$	$-1/14$	0	$5/14$	$5/2$
	0	0	$\dfrac{14P - 15}{14}$	$-P$	$\dfrac{56P - 9}{14}$	$\dfrac{-28P + 15}{14}$	0	$\dfrac{-70P + 9}{14}$	$\dfrac{56P + 33}{2}$

$\Rightarrow$ marks the second row.

In Tableau 3, x_5 has the largest positive difference coefficient, so x_5 is increased until it reaches the limiting value of 7 beyond which x_7 would become negative.   In Tableau 4 all difference coefficients are negative, so no further reduction in the objective function is possible.   The solution is

$$x_1 = 1 \quad \text{and} \quad x_2 = 5$$

at which point $y = 21$.

Tableau 4   Example 11-2

x_1	x_2	$x_3 = 0$	$x_4 = 0$	x_5	$x_6 = 0$	$x_7 = 0$	$x_8 = 0$	
1	0	$-13/56$	$1/56$	0	$13/56$	$-1/56$	0	1
0	0	$1/4$	$-1/4$	1	$-1/4$	$1/4$	-1	7
0	1	$9/56$	$-5/56$	0	$-9/56$	$5/56$	0	5
0	0	$-51/56$	$-9/56$	0	$-P + \dfrac{51}{56}$	$-P + \dfrac{9}{51}$	$-P$	21

11-16   Review of minimization calculation   Now that the minimization process in Example 11-2 has been completed, a reexamination of the successive tableaus in the problem will present a more complete picture of the operation.   The introduction of the artificial variables in the objective function and the constraint equations permits a temporary violation of the constraints, but only at the expense of an enormously large value of the objective function. The solution will certainly not be a satisfactory one until all of the P terms are removed from the expression for the objective function.

In contrast to the maximization problem, such as Example 11-1, where the constraints were solid walls that could not be surmounted, in the minimization problem the constraints are stiff rubber bands that can be violated temporarily, but with a severe penalty in the magnitude of the objective function.

Looking first at Tableau 1, the position represented in Fig. 11-9 is the origin, because $x_1 = x_2 = 0$. The slack variables x_3, x_4, and x_5 are also zero, but the artificial variables have nonzero values: $x_6 = 10$, $x_7 = 74$, and $x_8 = 9$. The value of the objective function is $93P$, which is prohibitively large.

In Tableau 2 the nonzero values of variables are $x_2 = 3$, $x_6 = 7$, and $x_7 = 35$. In moving from point 1 to point 2 in Fig. 11-9, the magnitude of the objective function drops from $93P$ to $42P + 9$. In other words, the pure numerical value increases by 9, but the staggering P term decreases from $93P$ to $42P$.

The next shift is to point 3, represented by Tableau 3 where the nonzero values are $x_1 = \frac{3}{2}$, $x_2 = \frac{5}{2}$, and $x_7 = 28$. The magnitude of the objective function is $(56P + 33)/2$ showing a continued increase of the numerical portion and a decrease of the P coefficient. At this point x_3 and x_6 are zero as are x_5 and x_8.

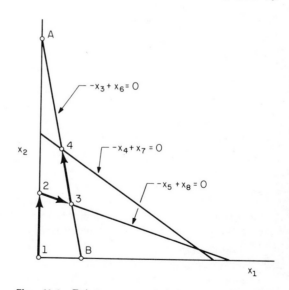

Fig. 11-9 Points represented by successive tableaux in Example 11-2.

Point 3 requires a positive value of the $-x_4 + x_7$ combination which is achieved by a zero value of x_4 and a positive value of x_7.

The final move is from point 3 to point 4 where the nonzero values are $x_1 = 1$, $x_2 = 5$, and the slack variable $x_5 = 7$. All of the artificial variables are now zero, so the objective function is free of P terms and has the value of 21. There are no remaining positive difference coefficients, so the minimum value has been reached.

The minimization process, then, consisted of starting at the origin, which was made permissible but highly uncomfortable by the introduction of artificial variables, and then moving through the thicket of constraint lines until finally seeing the light of day at point 4. In this problem, the first feasible point encountered was also the minimum, but it is possible that the process will move around several corners after it breaks into the feasible region.

One further comment should be made regarding Tableau 3 where, following the decision to program x_5, the limitation for increasing x_5 set by the first equation was 21, by the second equation 7, and by the third equation -7. To be strictly consistent, we would have chosen the third as the controlling equation. Geometrically, in Fig. 11-9, the test to find the controlling equation consists of moving along the $-x_3 + x_6 = 0$ line. Beyond point A, where $x_5 = 21$, the value of x_1 would go negative. Beyond 4 where $x_5 = 7$, x_7 would go negative. The limitation of $x_5 = -7$ is represented by point B which would represent a back step, so the limitation with a negative value is ignored.

11-17 Further topics in linear programming This chapter is an introduction to linear programming and considers only the straightforward maximization and minimization situations. Many practical problems require extensions into cases where the constraints are mixed such that they include both greater-than and less-than, as well as equality constraints. Troublemakers can appear due to what is called *degeneracy* and also *redundancy*.

A valuable contribution that linear programming can make is sensitivity analysis where, following the solution of the optimization, the individual coefficients in the constraint equations or the objective functions are analyzed to determine what influence a slight change in the coefficient would have on the optimal value. An application of sensitivity analysis might be to determine what the influence on the peak operating effectiveness of a processing plant might be if one certain heat exchanger could be enlarged.

Hopefully, the study of this chapter will permit the reader to recognize a linear-programming situation when it arises, will permit him to work straightforward linear-programming problems, and will encourage him to further study when encountering a more difficult linear-programming optimization.

SOME INTRODUCTORY TEXTS IN LINEAR PROGRAMMING

Charnes, A., W. W. Cooper, and A. Henderson: "An Introduction to Linear Programming," John Wiley & Sons, Inc., New York, 1953.
Dano, S.: "Linear Programming in Industry," 3rd ed., Springer Publishing Co., Inc., New York, 1965.
Dantzig, G. B.: "Linear Programming and Extension," Princeton University Press, Princeton, N.J., 1963.
Garvin, W. W.: "Introduction to Linear Programming," McGraw-Hill Book Company, New York, 1960.
Greenwald, D. U.: "Linear Programming," The Ronald Press Company, New York, 1957.
Hadley, G.: "Linear Programming," Addison-Wesley Publishing Company, Inc., Reading, Mass., 1962.
Llewellyn, R. W.: "Linear Programming," by Holt, Rinehart and Winston, Inc., New York, 1964.

PROBLEMS

11-1. During a summer session you enroll in two courses, psychology for four hours and engineering for three hours. You would like to use your time in such a way as to amass the largest number of grade points (numerical grade times the number of hours). From consultation with students who have taken these courses previously, you have established that the possible grades are functions of the time spent:

$$G_e = \frac{1}{4} x_1$$

and

$$G_p = \frac{1}{7} x_2$$

where G is the numerical grade (5.0 = A, 4.0 = B, etc.)

x_1 = number of hours per week spent outside of class studying engineering

x_2 = number of hours per week spent outside of class studying psychology

The total number of hours available for outside study per week cannot exceed 44. Furthermore, you can stomach the studying for these two courses in a combination such that

$$x_1 + 0.5x_2 \leq 30$$

You wish to obtain grades no lower than C, and observe also that hours spent studying beyond that necessary to earn an A would be fruitless.

By drawing all constraint lines and lines of constant grade points on a graph with coordinates of x_1 and x_2, determine the combination of study hours

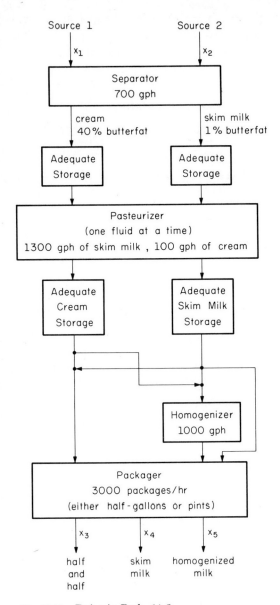

Fig. 11-10   Dairy in Prob. 11-2.

that result in the accumulation of the largest number of grade points. What grades would you obtain, and how many grade points would you acquire?
Ans.: B in both courses.

11-2. A dairy operating on the flow diagram shown in Fig. 11-10 can buy raw milk from either or both of two sources and can produce skim milk, homogenized milk, and half-and-half cream. The costs and butterfat contents of the sources and products are:

Item	Designation	Sale or purchase cost	Butterfat content, % by volume
Source 1	x_1 gal/day	$0.42 per gal	4.0
Source 2	x_2 gal/day	0.46 per gal	4.5
Half-and-half	x_3 gal/day	0.24 per pt	≥ 10.0
Skim milk	x_4 gal/day	0.29 per half gal	≥ 1.0
Homogenized milk	x_5 gal/day	0.33 per half gal	≥ 3.0

The daily quantities of sources and products are to be determined such that the plant operates with maximum profit. All units in the dairy can operate for a maximum of 8 hr/day.

(a) Write the equation for the profit in terms of x variables. Write the constraint equations.

Ans.: Profit $= -0.42x_1 - 0.46x_2 + 1.92x_3 + 0.58x_4 + 0.66x_5$

Separator: $\quad x_1 + x_2 \leq 5,600$

Pasteurizer: $\quad x_1 + 1.08x_2 \leq 5,408$

Homogenizer: $\quad x_5 \leq 8,000$

Packager: $\quad 4x_3 + x_4 + x_5 \leq 12,000$

Butterfat balance: $\quad -0.04x_1 - 0.045x_2 + 0.1x_3 + 0.01x_4 + 0.03x_5 \leq 0$

Total mass: $\quad -x_1 - x_2 + x_3 + x_4 + x_5 \leq 0$

There are 11 variables in the 6 constraint equations which means that any 5 variables could be set equal to zero and the remaining matrix solved for the other 6. The number of possible ways of choosing these five zero variables is

$$\frac{11!}{6!\,5!} = 462$$

These 462 possible solutions are distributed as follows:

196 cases are nonsoluble because the resulting equations are nonindependent
238 solutions have negative values of one or more variables
28 are valid solutions

The results of the 28 valid solutions are shown in Table 11-2 with slack variables designated x_6 through x_{11}.

Table 11-2   Solutions in dairy of Prob. 11-2

No.	Source 1 x_1	Source 2 x_2	Half and half x_3	Skim x_4	Homogenized x_5	Separator x_6	Pasteurizer x_7	Homogenizer x_8	Packager x_9	Butterfat balance x_{10}	Total mass balance x_{11}	Profit
1	*	*	*	*	*	5,600	10,400	8,000	12,000	0	0	0
2	*	*	*	*	0	5,600	10,400	8,000	12,000	0	*	0
3	*	*	*	0	*	5,600	10,400	8,000	12,000	*	0	0
4	*	*	*	0	*	5,600	10,400	8,000	12,000	0	*	0
5	*	*	*	0	0	5,600	10,400	8,000	12,000	*	*	0
6	*	*	0	*	*	5,600	10,400	8,000	12,000	0	*	0
7	*	*	0	*	0	5,600	10,400	8,000	12,000	*	*	0
8	*	*	0	0	*	5,600	10,400	8,000	12,000	*	*	0
9	*	5,007	*	*	*	593	*	8,000	12,000	225	5,007	−2,303
10	*	0	*	*	*	5,600	10,400	8,000	12,000	*	0	0
11	*	5,007	*	*	5,007	593	*	2,993	6,993	75	*	1,001
12	*	0	*	*	0	5,600	10,400	8,000	12,000	*	*	0
13	*	5,007	*	5,007	*	593	*	8,000	6,993	175	*	601
14	*	0	*	0	*	5,600	10,400	8,000	12,000	*	*	0
15	*	5,007	2,253	*	*	593	*	8,000	2,987	*	2,754	2,023
16	*	5,007	1,073	*	3,934	593	*	4,066	3,774	*	*	2,353
17	*	5,007	1,947	3,060	*	593	*	8,000	1,151	*	0	3,210
18	5,408	*	*	*	*	192	*	8,000	12,000	216	5,408	−2,271
19	0	*	*	*	*	5,600	10,400	8,000	12,000	*	0	0
20	0	*	*	*	*	5,600	10,400	8,000	12,000	0	*	0
21	5,408	*	*	*	5,408	192	*	2,592	6,592	54	*	1,298
22	0	*	*	*	0	5,600	10,400	8,000	12,000	*	*	0
23	5,408	*	*	5,408	*	192	*	8,000	6,592	162	*	865
24	0	*	*	0	*	5,600	10,400	8,000	12,000	*	*	0
25	5,408	*	2,163	*	*	192	*	8,000	3,347	*	3,245	1,882
26	0	*	0	*	*	5,600	10,400	8,000	12,000	*	*	0
27	5,408	*	773	*	4,636	192	*	3,364	4,274	*	*	2,271
28	5,408	*	1,803	3,605	*	192	*	8,000	1,184	*	*	3,281

* Indicates variables set equal to zero.

The following questions pertain to those solutions:

On the most profitable solution:

(b) Which is the most profitable solution?

(c) What are the daily quantities of sources and products?

(d) If you were to enlarge one component in order to increase the capacity of the plant, which one would it be?

(e) How many hours of the day is the separator idle?

On other solutions:

(f) In which solution(s) is the butterfat content in the product higher than necessary?

(g) In which solution(s) is milk being dumped?

(h) In what solution is the packager most heavily loaded?

(i) Physically we know that it is possible to buy some of each source and manufacture some of each product. Why is this condition not represented in any of the solutions?

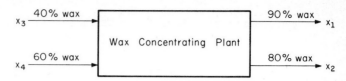

Fig. 11-11 Wax concentrating plant in Prob. 11-3.

(*j*) One of the many situations where the equations were not independent was where x_2, x_3, x_5, x_8, and x_9 were set equal to zero. Write the remaining equations and show why the program found the equations to be dependent.

11-3. A wax concentrating plant, as shown in Fig. 11-11, receives feedstock with a low concentration of wax and refines it into a product with a high concentration of wax. The selling prices of the products are x_1, $8 per hundred pounds; and x_2, $6 per hundred pounds; and the raw material costs are x_3, $1.5 per hundred pounds, and x_4, $3 per hundred pounds.

The plant operates under the following constraints:

(*a*) No more wax leaves the plant than enters.

(*b*) The receiving facilities of the plant are limited to no more than a total of 800 lb/hr.

(*c*) The packaging facilities can accommodate a maximum of 600 lb/hr of x_2 or 500 lb/hr of x_1 and can switch from one to the other with no loss of time.

If the operating cost of the plant is constant, use the simplex algorithm to determine the purchase and production plan that results in the maximum profit.

Ans.: $x_1 = 500$, $x_2 = 0$, $x_3 = 150$, and $x_4 = 650$ lb/hr
Profit $= \$18.25$ per hour

11-4. The optimization of the combined gas- and steam-turbine plant in Prob. 6-2 resulted in a linear objective function and three linear constraints. Use the simplex algorithm to determine the optimum values of q_1 and q_2. For ease in mathematical manipulation, use the following equations instead of those in Prob. 6-2.

Objective function:

$$q = q_1 + q_2$$

Subject to:

$$q_1 + 0.5q_2 \geq 7 \times 10^{10}$$
$$q_1 + 2.5q_2 \geq 12.5 \times 10^{10}$$
$$q_1 + 1.25q_2 \geq 10 \times 10^{10}$$

Ans.: $q_1 = 5 \times 10^{10}$ $\qquad q_2 = 4 \times 10^{10}$.

11-5. A chemical plant whose flow diagram is shown in Fig. 11-12 manufactures ammonia, hydrochloric acid, urea, ammonium carbonate, and ammonia chloride from carbon dioxide, nitrogen, hydrogen, and chlorine, The x values in Fig. 11-12 indicate flow rates in mol/hr.

The costs of the feed stocks are c_1, c_2, c_3, and c_4 \$/mol and the values of the products are p_5, p_6, p_7, and p_8 \$/mol where the subscript corresponds to

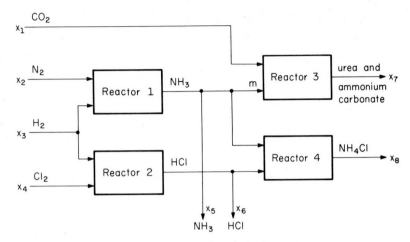

Fig. 11-12 Flow diagram of chemical plant in Prob. 11-5.

that of the x value. In reactor 3 the ratios of molal flow rates are $m = 3x_7$ and $x_1 = 2x_7$, and, in the other reactors, straightforward material balances apply. The capacity of reactor 1 is equal to or less than 2,000 mol/hr of NH_3 and the capacity of reactor 2 is equal to or less than 1,500 mol/hr of HCl.

    (*a*) Develop the expression for the profit.

    (*b*) Write the constraint equations for this plant.

Ans.: (*b*) $x_1 - 2x_7 = 0$

$$2x_2 - x_5 - 3x_7 - x_8 = 0$$
$$2x_3 - 3x_5 - x_6 - 9x_7 - 4x_8 = 0$$
$$2x_4 - x_6 - x_8 = 0$$
$$x_6 + x_8 \leq 1,500$$
$$x_5 + 3x_7 + x_8 \leq 2,000$$

Comprehensive Problems

This appendix presents some sample projects which apply principles studied in the text, such as economics, equation fitting, simulation, optimization, or a combination of these topics. The author uses such problems as these as projects accompanying the study of the text material and running as a part-time effort all term. In fact, some of the problems have carried over into additional terms, with one student or team of students picking up the work where the preceding group left off.

Engineering students become proficient in solving short problems—homework problems which require 45 min. Ironically, most professional engineering problems are long term, requiring weeks or months for completion. It is appropriate, then, for senior-level or graduate-student engineers to experience comprehensive problems which require discipline to maintain progress over a longer period of time. Also, at the beginning of any long-term project there is the period of deliberation on how even to start the problem—how to

find the handle.   Inexperienced engineers spend considerable time
wheel spinning and making false starts before focusing on a valid
solution.   Experience with comprehensive projects is the best means
of developing proficiency in thought and work habits.   A written
and also an oral report make good targets for completion and have
their own benefit as well.

The description of the projects that follow consists of a statement of the problem, which contains all or most of the required
data, and a discussion of some of the results or experiences of student
teams that have worked on the project.

A-1. Optimum temperature distribution in a multistage flash
      evaporation desalination plant.
A-2. Heat recovery from exhaust air with an ethylene glycol run-
      around system.
A-3. System for cooling and dehumidifying air for preserving grain.
A-4. Design of a fire-water grid.
A-5. Optimum thickness of insulation in refrigerated warehouses.
A-6. Simulation of a liquefied natural gas facility.

A-1 Optimum temperature distribution in a multistage flash evaporation desalination plant

Flash desalination   One of the methods for water desalination
is a distillation process using multistage flash evaporators, as shown
schematically in Fig. A-1.   Seawater flows first through heat

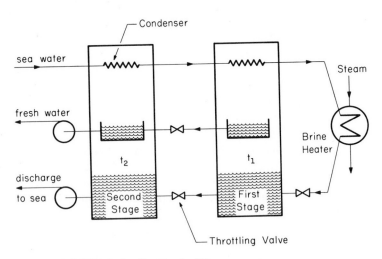

Fig. A-1 Multiflash desalination facility.

exchangers on which vapor condenses to form fresh water. After this preliminary heating of the seawater in the condensers, a steam heater elevates the temperature of the seawater to the maximum permitted by corrosion limitations. In passing through the first throttling valve some of the hot brine vaporizes, and this water vapor condenses on the tubes of the condenser and drops to the fresh-water collection pan. From the first stage the brine and the fresh water flow through pressure-reducing valves into the next stage.

Figure A-1 shows a two-stage plant, but commercial plants have many more stages. A Foster-Wheeler plant at San Diego, for example, has nine stages.[1]

The temperature in a stage equals the condensing temperature of the vapor which in turn is dictated by the amount of heat-transfer area in each stage of the condenser. The assignment is to specify the optimum number of stages and the number of square feet of heat-transfer area in each of the condensers for a minimum total cost over the 20-year life of the desalination plant, for a plant with a fresh-water capacity of 1 million gal/day.

Foster-Wheeler provided the following design data:

20-year life

5 percent interest on investment

Heat-exchanger cost—$6.50 per ft² for both condensers and brine heater

U value for both condensers and brine heater = 500 Btu/ (hr)(ft²)(°F)

Seawater temperature—60°F

Maximum permitted temperature of brine leaving the brine heater—250°F

Steam for the brine heater costs $0.30 per million Btu, and each additional flashing stage costs $10,000

Assume that the thermal properties of the brine are the same as those of pure water

Discussion Several student teams working on this project have had the following experiences. They found it advisable first to set up the thermodynamic and heat-transfer relations for a single-stage plant. An interesting observation is that the minimum stage temperature is the average of the seawater temperature and

[1] See W. A. Gardner, Desalination Test Module, *Heat Engineering*, Foster Wheeler Corporation, September–October, 1968; and W. A. Gardner, Concept for the Future, *Heat Engineering*, March–April, 1967.

the maximum brine temperature, thus $(60 + 250)/2 = 155°F$.   This minimum temperature occurs with a condenser of infinite area. The optimum condenser area results in a temperature that is a fraction of a degree above 155°F.

Moving to a two-stage plant, infinite area in the two condensers results in stage temperatures that again divide the 60 to 250°F range equally, but this time into thirds.   Thus, the minimum temperature of the first stage is 186.7°F, and with this first-stage temperature the minimum second-stage temperature is 123.3°F.

Turning to the optimization, a staged process such as this immediately suggests the method of dynamic programming.   The counterflow nature of the streams, however, prevents complete calculations at one stage before moving on to the next, and the elementary dynamic programming discussed in Chap. 9 is thus not adequate to provide the solution.   One group of students set up the objective function with a series of equality constraints and attempted the method of Lagrange multipliers.   In principle this method will work, but the equations became so long and detailed that this group never completely eliminated the algebraic errors that crept in.   The successful team of students used a lattice search, starting with areas resulting in temperatures that were within a degree or two of those indicated by the infinite-area temperature distribution.   Their optimal solution performed on a four-stage system consisted of an almost uniform distribution of condenser area among the stages.   As of this writing, no team has carried on the project beyond four stages to determine which is the optimum number of stages.

A-2   Heat recovery from exhaust air with an ethylene glycol run-around system

Introduction   Most large buildings are subject to ventilating requirements which necessitate bringing in outdoor air and exhausting an equal amount.   When the outdoor temperature is low, the cost of heating the outdoor air before introducing it to the space is appreciable.   This heating cost can be reduced by recovering some heat from the exhaust air.   One method of recovering heat is to place a finned-coil heat exchanger in the exhaust air stream, another in the outdoor air stream and to pump a fluid between the coils, as shown in Fig. A-2.   Water would be the first choice for a heat-transfer fluid, but to guard against freezeup at low outdoor temperature, an antifreeze such as ethylene glycol must be added.

If the system serves any purpose at all, it must save more

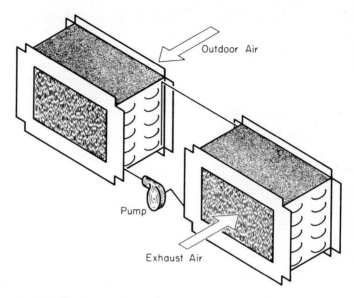

Fig. A-2   Heat-recovering system.

money in heating costs than it requires for its own amortization and operation. The first costs that are to be amortized include the costs of the coils, pump, piping, wiring, additional ductwork or revisions thereof, and additional cost of larger fans or motors, if needed. The operating costs include the power for the pump and additional fan power.

The potential savings should be evaluated for an optimum heat-recovery system. This optimum system consists of the most favorable combination of the following variables:

1. Length of the coil
2. Height (or number of tubes high) of coil
3. Number of rows of tubes deep (parallel to the path of air flow)
4. Fin spacing
5. Glycol flow rate

The qualitative effects of the previously mentioned five variables are as follows: increasing the length and number of tubes high increases the heat-transfer area, but also the cost. Furthermore, increasing the length and number of tubes high increases the cross-sectional area for air flow which reduces the velocity and decreases the air-side heat-transfer coefficient, but also decreases the power

required of the fan to force the air through the coil.   Increasing the
number of rows of tubes deep increases the heat-transfer area, but
increases the first cost of the coil as well as both the air and the
glycol pumping cost.   Spacing the fins closer together increases
the cost of the coil and the air-pressure drop, but also increases the
heat-transfer area.   Finally, a high flow rate of the ethylene glycol
increases the glycol side heat-transfer coefficient, but also increases
the pumping cost.

Further data and assumptions   The size of the optimum system
will pertain to a given air-flow rate, and 1,000 cfm has been chosen
for both the outdoor-air flow rate and the flow of exhaust air.   The
coil circuiting chosen is that of vertical headers feeding horizontal
tube circuits in parallel, as shown in Fig. A-2.   The flow of the
glycol through the U bends is counter to the flow of air.   Assume
pure counterflow in the coils, thus the equations that represent this
run-around loop are those used in Prob. 4-10.
Further assumptions are:

1. Tubes of $\frac{5}{8}$ in. outside diameter with an 0.017 in. wall thickness
2. Tube spacing $1\frac{5}{8}$ in. vertically and horizontally
3. Average outdoor temperature, 40°F for 250 days of operation
4. Ten-year life, 6 percent interest
5. Electricity cost—$0.015 per kWh
6. Fan pump and motor efficiencies—75 percent

The optimization   The net saving to be maximized is the difference between the reduction in heating cost and the annual cost
(first plus operating) of the recovery system.   The first cost to be
amortized includes that of the coils, the pump and motor (assumed
constant at $300), and the interconnecting piping (assumed con-
stant at $100).   No allowance for capital cost for the fan is pro-
vided, because fans already exist in the system and it is assumed
they would not have to be enlarged to overcome the additional
pressure drop of the heat-recovery coils.

Cost and performance equations   An equation for the first cost
of the coil that reflects the rough proportionality of the cost to the
weight of the coil is:

$$\text{Cost}(\$) = 0.00146[4.0 + (0.155)(NF + 16.0)W][L(NR - 1)]$$

where NF = coil fin spacing, fins/in.
      W = number of rows of tubes deep

L = tube length, in.

NR = number of layers of tubes high

The U value of the coil based on outside (air-side) area:

$$\frac{1}{U_0} = \frac{2.75}{V_a^{0.58}} + 0.0183 + \frac{1.975NF + 1.05}{166 V_{eg}^{0.8}}$$

where U_0 = U value, Btu/(hr)(ft²)(°F)

V_a = face velocity of the air through the coil, fpm

NF = fin spacing, fins/in.

V_{eg} = velocity of ethylene glycol in tubes, fps

(the resistance of the tubes and fins is 0.0183)

Air pressure drop:

$$PA = (0.0804 \times 10^{-5})(0.25W)(0.25 + 0.0625NF)(V_a^{1.725})$$

where PA = pressure drop of air, in. of water

Pressure drop of the ethylene glycol:

$$PEG = 1.495 \left(\frac{V_{eg}}{3}\right)^{17.5} [0.0039WL + 0.0875(W - 1) + 0.3]$$

where PEG = pressure drop of the ethylene glycol through the coil, ft of water

In addition, some allowance should be made for the interconnecting piping, 5 ft of water, for example.

Variables   The five variables for which optimal values are sought and the ranges of their values are

NF, fin spacing, fins/in.	8 to 14 by 2s
L, coil tube length, in.	12 to 144 by 6s
W, number of rows of tubes deep	2 to 10 by 2s
NR, number of rows of tubes high	2 to $\dfrac{L}{2.25}$ by 1s
V_{eg}, velocity of ethylene glycol, ft-sec	1 to 6 by 1s

A specific combination of the above variables should be optimum for a given rate of air flow and a given heating cost.   A heating cost somewhere between 0.1 to 0.5 cents per thousand Btu may be chosen.

Discussion   An optimization performed by a univariate search (as explained in Chap. 8) indicated the optimum values of three of the variables to be at their minimum quantities.   These were $NF = 8$, $V_{eg} = 1$, and $W = 2$.   Since the selection of ranges was arbitrary it would be advisable to investigate wider fin spacing, lower glycol velocities, and one-row-deep coils.

A-3   System for cooling and dehumidifying air for preserving grain

Introduction   A radical revision in the practice of harvesting corn in the midwest during the past decade has been brought about by widespread combining of the grain. Instead of permitting the grain to dry on the ear in the field and then harvesting it by separate picking and shelling operations, farmers now use one machine to perform the entire operation. The combine is a heavy machine which cannot operate in wet fields late in the year, so the harvest must take place earlier in the year while the grain is still moist.

Heated air most frequently serves as the means for drying the air, but hot air mars the quality of the corn for many eventual uses. Unheated outdoor air forced through the grain bed will also dry the corn. The third possible treatment of the air before delivering it to the grain is to dehumidify it as shown in Fig. A-3. The conditioning system in Fig. A-3 is identical to the cycle used in a home appliance dehumidifier and is only one of several possible arrangements. Instead of using the condenser heat to reheat the air, this condenser heat can be rejected to the ambient. Another version of the air circuit is to recirculate some or all of the exhaust air. The basic objective is to cool and/or dry the grain—both of which conditions enhance the storage time of the grain before any spoilage occurs, as is illustrated by Fig. A-4.

Assignment   For a 20,000-bu storage facility with an air-flow rate of 0.5 cfm/bu, compute the conditions of air entering the grain bed for a variety of outdoor-air conditions. The compressor characteristics are the same as those listed in Prob. 5-4. The

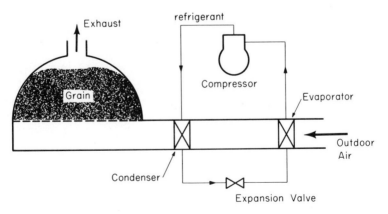

Fig. A-3   Grain-conditioning system.

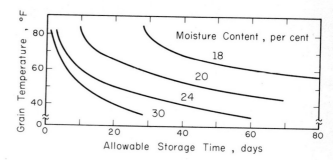

Fig. A-4 Allowable storage time before spoilage of the grain. (*Data from USDA Grain Storage Research Laboratory, Ames, Iowa.*)

product of UA for both the condenser and evaporator is 45,000 Btu/(hr)(ft²)(°F).

Discussion A student team that attacked this problem selected as their first objective the simulation of the cooling plant, assuming only sensible cooling (no dehumidification) at the evaporator. This step is a logical extension of Prob. 5-4, adding another equation and another unknown—the outlet-air temperature from the evaporator. As of this date, dehumidification at the evaporator coil has not been incorporated in the simulation program. With no dehumidification the net effect of the refrigeration plant is to raise the temperature of the air by adding the heat equivalent of the power required by the compressor. The extension to include evaporator dehumidification requires introduction of the properties of moist air and the inclusion of mass transfer as well as heat transfer at the evaporator.

A-4 Design of a fire-water grid

Introduction Refineries and other chemical plants that process flammable substances must provide elaborate measures to prevent and fight fires. Almost all such plants must be equipped with a fire-water distribution system that is generally located underground to supply hydrants throughout the plant. Typically, the plant is subdivided into various areas, remote enough from one another that should a fire break out in one area it could be contained within that area. To serve the entire plant, the fire-water grid should be capable of providing a specified rate of water flow to any one area at a time.

Two challenges appear before the designer. For a given pressure at the outlet of the pump:

1. Select the pipe sizes so that the flow through the grid will supply the hydrants that surround a plant area with the required rate of water flow.
2. Design the minimum-cost piping system that meets the requirements of (1).

Most industrial designers of fire-water grids are satisfied to achieve task 1, and even this assignment is an interesting experience. Some firms use specially designed electrical analogs (specially designed because the fluid-flow conductor does not follow Ohm's law) in combination with a cut-and-try method of enlarging pipe sizes until each area individually could be blanketed with specified water flow.

Objective The goal of this project is to design (select pipe sizes) a fire-water grid having minimum first cost when the following conditions are specified:

1. Geometric layout of areas
2. Location of hydrants
3. Water pressure at pump output
4. Minimum flow rate required by each operating area

Although in actual plants there may be a dozen operating areas with hydrants distributed along the pipes, a suggested grid for an initial solution of this type of problem is shown in Fig. A-5.

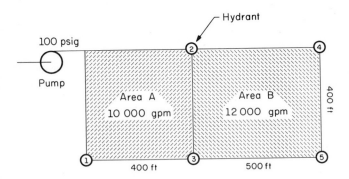

Fig. A-5   Fire-water grid.

Operating and cost data When the pump supply pressure is 100 psig, the grid must supply a minimum of 10,000 gpm to operating area A or 12,000 gpm to operating area B, but not simultaneously. Area A is served by hydrants 1, 2, and 3, while area B is served by hydrants 2, 3, 4, and 5. The flow characteristics of the hydrants are:

$$Q \text{ (gpm)} = 450 \sqrt{p \text{ (psig)}}$$

The installed cost of the pipe is $2 per ft length per inch diameter. The trenching cost is $20 per ft, but since it is independent of the pipe size it is a constant and does not affect the optimum pipe-size selection. Neglect operating cost because the number of hours of pump operation, hopefully, will be negligible.

Discussion Finding the optimum grid design actually resolves to a separate optimization, focusing on one operating area at a time. In the grid of Fig. A-5, for example, the optimum design if area A demands water will be found separately from the design if area B is in operation. The adequate system which serves area A alone will probably be inadequate for area B alone, so in this case the optimum design that serves area B should be checked to see whether it is adequate for area A.

The experience of several student teams working on this problem is that the optimization problem reduces to one involving equality constraints and thus is soluble by the method of Lagrange multipliers. One difficulty that emerges is the amount of algebra which is unique to each grid design. The possibility of mathematical errors when moving from the simplified grid to realistic grids suggests the advisability of a master computer program into which the specific conditions of the existing plan could be inserted. Since the piping cost of these grids often is of the order of hundreds of thousands of dollars, some engineering time is a good investment if, for example, 10 percent of the first cost could be saved.

Another refinement that adds more constraints is that there should be a minimum pressure at the hydrant—perhaps 50 psig. The reason for this minimum is that adequate pressure must be available to send a stream of water at least into the center of the area, and perhaps further, in order to provide some overlap of water coverage.

A-5 Optimum thickness of insulation in refrigerated warehouses

Introduction Deciding upon the optimum insulation thickness is a classic engineering problem. When heat transfer into or out of a controlled-temperature space results in cooling or heating cost, application of insulation will reduce these costs.   Each additional inch of insulation, however, is progressively less effective, and the point is reached where additional thickness of insulation is not compensated by the reduction in operating cost.

A large food company suggested that some of the standard thicknesses of insulation used in refrigerated warehouses should be restudied.   Expanded foam insulation has replaced cork as insulating material, but the conventional thicknesses applicable to cork are still being used.   This practice persists even though the insulating value of the new insulation is better than cork, and the installed cost of refrigeration equipment has increased during the past decade.

This study, then, is an old idea, but includes some of the realistic economic considerations of a practical engineering study.

Cost and engineering data   A building of 10,000 ft² floor area with 16-ft-high walls would be of reasonable dimensions for a refrigerated warehouse.   The warehouse temperature is 0°F and the average outdoor temperature during the year will be between 40 and 60°F, depending upon the climate of the area.

The expanded foam insulation is available in 2-, 3-, 4-, and 5-in. thicknesses.   To develop thicknesses greater than 5 in., one or two additional layers of insulating board are required.   The insulating value of the structural material (concrete, for example) should also be considered.   Additional data are:

Cost of insulation	$0.09 per board ft (1 in. × 1 ft²)
Cost of installing insulation	$0.50 per ft² to apply first layer
	$0.35 per ft² to apply second layer
	$0.50 per ft² to finish surface
First cost of refrigeration equipment	$1,500 per ton of refrigeration
Power requirements by refrigeration plant	2 hp/ton
Electrical power cost	depends upon local rates—usually between $0.01 and $0.02 per kWh
Sixteen-year life	8 percent interest
Write off on equipment for tax purposes, straight-line depreciation	10 years
Federal income tax	50 percent
Conductivity of insulation	$k = 0.25$ Btu-in./(hr)(ft²)(°F)

Discussion   Two student teams have worked on this project. Both made present-worth analyses writing computer programs performing an exhaustive search through all the thicknesses of insulation available.

The first group submitted their report recommending optimum insulation thicknesses from 20 to 30 in.! This experience was a good opportunity to emphasize the need for always questioning the reasonableness of calculated results. Soon we discovered that when the students found the value of conductivity of the insulating board in the insulating manufacturer's brochure to be 0.25, they did not notice that the units were Btu-in./(hr)(ft²)(°F) rather than Btu/(hr)(ft)(°F) as they have been accustomed to using in their heat-transfer class.

The second group found that a 5-in. thickness was generally the optimum, except at very high power costs where the optimum jumped up to 10 in. The reason that the other integer thicknesses rarely appeared is because of the large percentage of the cost associated with the installation of the insulation and the smaller effect of the cost of the insulation itself.

A-6   Simulation of a liquefied natural gas facility

Introduction   The gas utility companies are "distribution" companies and they purchase natural gas, which is predominantly methane, from transmission companies. The transmission companies assess demand charges, so it is often to the advantage of the distribution company to buy gas from the transmission line during periods of low demand and store the gas nearby for use during the peak winter periods.

The two principal storage concepts used most widely are:

1. Underground storage as a high-pressure gas
2. Storage above ground in liquid form

The geological formations in the area often do not permit underground storage, so the storage must then be above ground. The advantage of storage in liquid form over storage as a gas is in the reduced storage volume. Six hundred times as much gas can be stored in liquid form as in gaseous form at atmospheric pressure. Of course, the liquefied natural gas (LNG) is at a low temperature, −250°F, so there is a cost associated with the liquefaction, maintaining the low temperature and subsequent revaporization. Nevertheless, LNG plants are coming into wide use both in domestic service and for liquefaction of gas prior to transoceanic shipping.

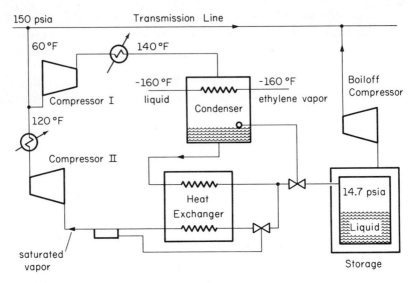

Fig. A-6   LNG plant.

The LNG plant   Figure A-6 shows a schematic diagram of a simplified LNG plant.   The actual plant may utilize a propane refrigeration plant for the first stage of cooling, an ethylene plant for the second stage, and several additional heat exchangers to improve the cycle efficiency.   In the cycle of Fig. A-6, gas from the transmission line enters the liquefaction plant at 150 psia and 60°F. Following compression and cooling by ambient air to 140°F, the methane condenses in a heat exchanger which is a methane condenser and an ethylene evaporator.   The remainder of the ethylene refrigeration plant is not under consideration in this study.   A liquid-level valve in the condenser assures that saturated liquid leaves the condenser and enters the heat exchanger where the liquid methane is subcooled.   A fraction of the liquid methane leaving the heat exchanger expands through a throttling valve and evaporates on the other side of the heat exchanger, leaving as saturated vapor.   This saturated vapor is compressed back to 150 psia, cooled to 120°F, and mixes with the incoming feed for recycling through the liquefaction plant.   Returning to the outlet liquid stream from the heat exchanger, the fraction that is not recycled through the heat exchanger passes through a throttling valve down to atmospheric pressure.   The low-temperature liquid is stored in an insulated vessel, and any methane that flashes to vapor during the expansion is compressed back to 150 psia for return to the transmission line. The boil-off compressor maintains atmospheric pressure in the storage vessel by also compressing the vapor which boils due to heat leakage into the tank.

Objective and data Simulate this LNG system and compute the temperatures, pressures, and flow rates at all points in the cycle. The performances of the two compressors are given by the equations:

Compressor I:

$$\frac{p_2}{p_1} = 7.5 - 0.00369 \left[\frac{w \text{ (lb/hr)} \sqrt{T_{inlet}(°R)}}{p_1 \text{ (psia)}} \right]$$

where p_1 and p_2 are the inlet and discharge pressures, respectively, psia.

Compressor II:

$$\frac{p_2}{p_1} = 5.6 - 0.0035 \left[\frac{w \text{ (lb/hr)} \sqrt{T_{inlet}(°R)}}{p_1 \text{ (psia)}} \right]$$

The boil-off compressor has adequate capacity to maintain 14.7 psia in the storage vessel.

The UA of the condenser = 120,000 Btu/(hr)(°F). Even though methane enters the condenser superheated, assume in the heat-transfer calculation that the temperature of the methane is its saturation temperature throughout.

The UA of the heat exchanger = 10,500 Btu/(hr)(°F).

Some equations for methane properties that are reasonably accurate follow.

Enthalpy of saturated liquid, Btu/lb:

$$h_f = 323.2 + 1.56t + 0.00145t^2$$

where t = liquid temperature, °F

Enthalpy of saturated vapor, Btu/lb:

$$h_g = 169.4 - 0.961t - 0.00269t^2$$

where t = vapor temperature, °F

Relation of temperature and pressure at saturated conditions:

$$\ln p = -\frac{1,852}{T} + 11.89$$

where p = pressure, psia
T = temperature, °R

Discussion   Using the system-simulation procedures of Chap. 5 with the aid of a computer program, a student team found that the program converged after about five iterations.   Several key results were:

Pressure leaving compressor I   = 440 psia

Pressure entering compressor II = 55 psia

Liquefaction rate   = 3,950 lb/hr

Index